REVIEW OF THE ARMY'S TECHNICAL GUIDES ON ASSESSING AND MANAGING CHEMICAL HAZARDS TO DEPLOYED PERSONNEL

Subcommittee on the Toxicological Risks to Deployed Military Personnel

Committee on Toxicology

Board on Environmental Studies and Toxicology

Division on Earth and Life Studies

NATIONAL RESEARCH COUNCIL
OF THE NATIONAL ACADEMIES

THE NATIONAL ACADEMIES PRESS
Washington, D.C.
www.nap.edu

THE NATIONAL ACADEMIES PRESS **500 Fifth Street, NW** **Washington, D.C. 20001**

NOTICE: The project that is the subject of this report was approved by the Governing Board of the National Research Council, whose members are drawn from the councils of the National Academy of Sciences, the National Academy of Engineering, and the Institute of Medicine. The members of the committee responsible for the report were chosen for their special competences and with regard for appropriate balance.

This project was supported by Contract No. DAMD 17-99-C-9049 between the National Academy of Sciences and the U.S. Department of Defense. Any opinions, findings, conclusions, or recommendations expressed in this publication are those of the author(s) and do not necessarily reflect the view of the organizations or agencies that provided support for this project.

International Standard Book Number 0-309-09221-3 (Book)
International Standard Book Number 0-309-53239-6 (PDF)

Additional copies of this report are available from:

The National Academies Press
500 Fifth Street, NW
Box 285
Washington, DC 20055

800-624-6242
202-334-3313 (in the Washington metropolitan area)
http://www.nap.edu

Printed in the United States of America

THE NATIONAL ACADEMIES

Advisers to the Nation on Science, Engineering, and Medicine

The **National Academy of Sciences** is a private, nonprofit, self-perpetuating society of distinguished scholars engaged in scientific and engineering research, dedicated to the furtherance of science and technology and to their use for the general welfare. Upon the authority of the charter granted to it by the Congress in 1863, the Academy has a mandate that requires it to advise the federal government on scientific and technical matters. Dr. Bruce M. Alberts is president of the National Academy of Sciences.

The **National Academy of Engineering** was established in 1964, under the charter of the National Academy of Sciences, as a parallel organization of outstanding engineers. It is autonomous in its administration and in the selection of its members, sharing with the National Academy of Sciences the responsibility for advising the federal government. The National Academy of Engineering also sponsors engineering programs aimed at meeting national needs, encourages education and research, and recognizes the superior achievements of engineers. Dr. Wm. A. Wulf is president of the National Academy of Engineering.

The **Institute of Medicine** was established in 1970 by the National Academy of Sciences to secure the services of eminent members of appropriate professions in the examination of policy matters pertaining to the health of the public. The Institute acts under the responsibility given to the National Academy of Sciences by its congressional charter to be an adviser to the federal government and, upon its own initiative, to identify issues of medical care, research, and education. Dr. Harvey V. Fineberg is president of the Institute of Medicine.

The **National Research Council** was organized by the National Academy of Sciences in 1916 to associate the broad community of science and technology with the Academy's purposes of furthering knowledge and advising the federal government. Functioning in accordance with general policies determined by the Academy, the Council has become the principal operating agency of both the National Academy of Sciences and the National Academy of Engineering in providing services to the government, the public, and the scientific and engineering communities. The Council is administered jointly by both Academies and the Institute of Medicine. Dr. Bruce M. Alberts and Dr. Wm. A. Wulf are chair and vice chair, respectively, of the National Research Council

www.national-academies.org

SUBCOMMITTEE ON TOXICOLOGICAL RISKS TO DEPLOYED MILITARY PERSONNEL

RICHARD J. BULL (*Chair*), Consultant, Richland, WA
EDWARD BISHOP, Parsons Corporation, Fairfax, VA
KENNETH T. BOGEN, Lawrence Livermore National Laboratory, Livermore, CA
BARBARA CALLAHAN, University Research Engineers and Associates, Grantham, NH
JUDITH GRAHAM, American Chemistry Council, Arlington, VA
DAVID MOORE, Battelle Eastern Science and Technology Center, Aberdeen, MD
DEBORAH IMEL NELSON, University of Oklahoma, Norman
CHARLES F. REINHARDT, Consultant, Chadds Ford, PA
ROSALIND A. SCHOOF, Integral Consulting, Inc., Mercer Island, WA
ROBERT G. TARDIFF, The Sapphire Group, Inc., Vienna, VA
NGA L. TRAN, Exponent, Inc., Washington, DC

Staff

SUSAN N.J. MARTEL, Project Director
KELLY CLARK, Editor
TAMARA DAWSON, Program Assistant

Sponsor: U.S. Department of Defense

COMMITTEE ON TOXICOLOGY

BAILUS WALKER, JR. *(Chair)*, Howard University Medical Center and American Public Health Association, Washington, DC
MELVIN E. ANDERSEN, CIIT-Centers for Health Research, Research Triangle Park, NC
EDWARD C. BISHOP, Parsons Corporation, Fairfax, VA
GARY P. CARLSON, Purdue University, West Lafayette, IN
JANICE E. CHAMBERS, Mississippi State University, Mississippi State
LEONARD CHIAZZE, JR., Georgetown University, Washington, DC
JUDITH A. GRAHAM, American Chemistry Council, Arlington, VA
SIDNEY GREEN, Howard University, Washington, DC
MERYL KAROL, University of Pittsburgh, Pittsburgh, PA
STEPHEN U. LESTER, Center for Health Environment and Justice, Falls Church, VA
DAVID H. MOORE, Battelle Memorial Institute, Bel Air, MD
CALVIN C. WILLHITE, Department of Toxic Substances, State of California, Berkeley
GERALD WOGAN, Massachusetts Institute of Technology, Cambridge

Staff

KULBIR S. BAKSHI, Program Director
ROBERTA M. WEDGE, Senior Program Officer for Risk Analysis
SUSAN N. J. MARTEL, Senior Program Officer
ELLEN K. MANTUS, Senior Program Officer
KELLY CLARK, Assistant Editor
AIDA NEEL, Senior Program Assistant
TAMARA DAWSON, Program Assistant

Board on Environmental Studies and Toxicology[1]

---

[1]This study was planned, overseen, and supported by the Board on Environmental Studies and Toxicology.

OTHER REPORTS OF THE BOARD ON ENVIRONMENTAL STUDIES AND TOXICOLOGY

Air Quality Management in the United States (2004)
Endangered and Threatened Species of the Platte River (2004)
Atlantic Salmon in Maine (2004)
Endangered and Threatened Fishes in the Klamath River Basin (2004)
Cumulative Environmental Effects of Alaska North Slope Oil and Gas Development (2003)
Estimating the Public Health Benefits of Proposed Air Pollution Regulations (2002)
Biosolids Applied to Land: Advancing Standards and Practices (2002)
Ecological Dynamics on Yellowstone's Northern Range (2002)
The Airliner Cabin Environment and Health of Passengers and Crew (2002)
Arsenic in Drinking Water: 2001 Update (2001)
Evaluating Vehicle Emissions Inspection and Maintenance Programs (2001)
Compensating for Wetland Losses Under the Clean Water Act (2001)
A Risk-Management Strategy for PCB-Contaminated Sediments (2001)
Acute Exposure Guideline Levels for Selected Airborne Chemicals (4 volumes, 2000-2003)
Toxicological Effects of Methylmercury (2000)
Strengthening Science at the U.S. Environmental Protection Agency (2000)
Scientific Frontiers in Developmental Toxicology and Risk Assessment (2000)
Ecological Indicators for the Nation (2000)
Modeling Mobile-Source Emissions (2000)
Waste Incineration and Public Health (1999)
Hormonally Active Agents in the Environment (1999)
Research Priorities for Airborne Particulate Matter (4 volumes, 1998-2003)
Ozone-Forming Potential of Reformulated Gasoline (1999)
The National Research Council's Committee on Toxicology: The First 50 Years (1997)
Carcinogens and Anticarcinogens in the Human Diet (1996)
Upstream: Salmon and Society in the Pacific Northwest (1996)
Science and the Endangered Species Act (1995)
Wetlands: Characteristics and Boundaries (1995)
Biologic Markers (5 volumes, 1989-1995)
Review of EPA's Environmental Monitoring and Assessment Program (3 volumes, 1994-1995)
Science and Judgment in Risk Assessment (1994)
Pesticides in the Diets of Infants and Children (1993)
Dolphins and the Tuna Industry (1992)
Science and the National Parks (1992)
Human Exposure Assessment for Airborne Pollutants (1991)
Rethinking the Ozone Problem in Urban and Regional Air Pollution (1991)
Decline of the Sea Turtles (1990)

Copies of these reports may be ordered from the National Academies Press
(800) 624-6242 or (202) 334-3313
www.nap.edu

OTHER REPORTS OF THE COMMITTEE ON TOXICOLOGY

Spacecraft Water Exposure Guidelines for Selected Contaminants, Volume 1 (2004)
Toxicologic Assessment of Jet-Propulsion Fuel 8 (2003)
Review of Submarine Escape Action Levels for Selected Chemicals (2002)
Standing Operating Procedures for Developing Acute Exposure Guideline Levels for Hazardous Chemicals (2001)
Evaluating Chemical and Other Agent Exposures for Reproductive and Developmental Toxicity (2001)
Acute Exposure Guideline Levels for Selected Airborne Contaminants, Volume 1 (2000), Volume 2 (2002), Volume 3 (2003), Volume 4 (2004)
Review of the US Navy's Human Health Risk Assessment of the Naval Air Facility at Atsugi, Japan (2000)
Methods for Developing Spacecraft Water Exposure Guidelines (2000)
Review of the U.S. Navy Environmental Health Center's Health-Hazard Assessment Process (2000)
Review of the U.S. Navy's Exposure Standard for Manufactured Vitreous Fibers (2000)
Re-Evaluation of Drinking-Water Guidelines for Diisopropyl Methylphosphonate (2000)
Submarine Exposure Guidance Levels for Selected Hydrofluorocarbons: HFC-236fa, HFC-23, and HFC-404a (2000)
Review of the U.S. Army's Health Risk Assessments for Oral Exposure to Six Chemical-Warfare Agents (1999)
Toxicity of Military Smokes and Obscurants, Volume 1(1997), Volume 2 (1999), Volume 3 (1999)
Assessment of Exposure-Response Functions for Rocket-Emission Toxicants (1998)
Toxicity of Alternatives to Chlorofluorocarbons: HFC-134a and HCFC-123 (1996)
Permissible Exposure Levels for Selected Military Fuel Vapors (1996)
Spacecraft Maximum Allowable Concentrations for Selected Airborne Contaminants, Volume 1 (1994), Volume 2 (1996), Volume 3 (1996), Volume 4 (2000)

Preface

Military deployments include a spectrum of military activities ranging from peace-keeping, humanitarian, and nation-building missions to combat. In deployment situations, commanders must consider and balance a variety of hazards to the mission and to the health of their troops. To facilitate consideration of chemical threats in the decision-making process for mission planning, the U.S. Army has developed two technical guides (Technical Guide 230 and Technical Guide 248) and one reference guide (Reference Document 230) that outline a process by which chemical hazards can be characterized in terms of their health risks and categorized in terms of their impact on the mission (e.g., mission capable, combat ineffective). A key element of the guidance was the establishment of military exposure guidelines (MEGs) for air, water, and soil that are to be used for assessing the significance of field exposures to chemical hazards during deployment.

In this report, the National Research Council's (NRC) Subcommittee on Toxicological Risks to Deployed Military Personnel evaluates the Army's three guidance documents for their scientific validity and adequacy in characterizing chemical risks for comparison with other health and operational risks. Specifically, the subcommittee evaluated the adequacy of the proposed MEGs for assessing risks to soldier health and missions, the methods and special military considerations that should be used in developing exposure guidelines, and the application of the guidelines.

This report has been reviewed in draft form by individuals chosen for their diverse perspectives and technical expertise, in accordance with procedures approved by the NRC's Report Review Committee. The purpose of this independent review is to provide candid and critical comments that will assist the institution in making its published report as sound as possible and to ensure that the report meets institutional standards for objectivity, evidence, and responsiveness to the study charge. The review comments and draft manuscript remain confidential to protect the integrity of the deliberative process. We wish to thank the following individuals for their review of this report:

Germaine Buck, National Institute of Child Health and Human Development
Jeffrey Fisher, University of Georgia
Howard Kipen, University of Medicine and Dentistry of New Jersey
David Macys, University of Washington
Roger O. McClellan, Albuquerque, New Mexico
Lorenz Rhomberg, Gradient Corporation
Joseph Rodricks, ENVIRON International Corporation
Smita Siddhanti, EnDyna, Inc.
Palmer W. Taylor, University of California, San Diego

Although the reviewers listed above have provided many constructive comments and suggestions, they were not asked to endorse the conclusions or recommendations, nor did they see the final draft of the report before its release. The review of this report was overseen by Gilbert Omenn, University of Michigan, and Raymond Wymer, Oak Ridge, Tennessee. Appointed by the NRC, they were responsible for making certain that an independent examination of this report was carried out in accordance with institutional procedures and that all review comments were carefully considered. Responsibility for the final content of this report rests entirely with the authoring committee and the institution.

The subcommittee gratefully acknowledges the following individuals for making presentations and providing information to the subcommittee: LTC John Ciesla, U.S. Army Center for Health Promotion and Preventive Medicine (USACHPPM); Ellen Embry, Deputy Assistant Secretary of Defense for Force Health Protection; Robert Garrett, Armed Forces Medical Intelligence Center; Veronique Hauschild, USACHPPM; Jack Heller, USACHPPM; Joleen Mobley, USACHPPM; and Tony Pitrat, USACHPPM.

The subcommittee is also grateful for the assistance of the NRC staff in preparing this report. It particularly wishes to acknowledge the contributions of Susan Martel, project director, who coordinated the project and contributed to the subcommittee's report. Other staff members who contributed to this effort are Kulbir Bakshi, senior program officer for toxicology; Kelly Clark, assistant editor; Mirsada Karalic-Loncarevic, research associate; and Tamara Dawson, program assistant.

We would especially like to thank all the members of the subcommittee for their efforts throughout the development of this report.

Richard J. Bull, *Chair*
Subcommittee on Toxicological Risks to Deployed Military Personnel

Bailus Walker, *Chair*
Committee on Toxicology

Contents

ABBREVIATIONS .... *xvii*

SUMMARY *1*

1 INTRODUCTION *16*
Background, *16*
Guidance Documents, *21*
Statement of Task, *23*
The Subcommittee's Approach, *24*
References, *25*

2 REVIEW OF THE ARMY'S TECHNICAL GUIDANCE *27*
Earlier Academies Reports on Developing Reliable Comparative Risk Assessments for Deployments, *27*
The Army's Risk Assessment Guidance for Deployment, *30*
Recommendations, *45*
References, *46*

3 REVIEW OF KEY CONCEPTS, ASSUMPTIONS, AND DECISIONS MADE IN DEVELOPING TG-248, TG-230, AND RD-230 .... *47*
Use of Pre-Existing Exposure Guidelines, *47*
Acceptable Cancer Risk, *60*
Consideration of Immediate and Long-term Health Effects, *63*
Aggregate Exposure and Cumulative Risk, *65*
Exposure Assessment, *67*

Utility for Decision Makers, *70*
Recommendations, *71*
References, *72*

4 A NEW SET OF EXPOSURE GUIDELINES: CHEMICAL CASUALTY ESTIMATING GUIDELINES............... *76*
Introduction, *76*
Derivation of Chemical Casualty Estimating Guidelines, *78*
Application and Interpretation of CCEGs, *84*
Aggregate Exposure and Cumulative Risk, *87*
Recommendations, *87*
References, *89*

5 PROCESS FOR ESTABLISHING AND APPLYING MILITARY EXPOSURE GUIDELINES... *91*
Air Exposure Guidelines, *91*
Drinking Water Guidelines, *105*
Soil Exposure Guidelines, *110*
Application of MEGs, *119*
Recommendations, *125*
References, *127*

Appendix A. Errata, Inconsistencies, and Comments on Specific Aspects of TG-248, TG-230, and RD-230, *133*

Appendix B. Review of Acceptable Cancer Risk Levels, *137*

Appendix C. Example Use of Probits for Developing Chemical Casualty Estimating Guidelines, *145*

Appendix D. Critical Studies and Uncertainty Factors Used in Developing Acute Exposure Guideline Levels for Chemical Warfare Agents, *175*

Appendix E. Probabilistic Approach to Address Exposure to Multiple Chemicals for Course-of-Action Analysis, *181*

Appendix F. Biographical Information on the Subcommittee on Toxicological Risks to Deployed Military Personnel, *187*

Appendix G. Definitions, *193*

Abbreviations

ACGIH	American Conference of Governmental Industrial Hygienists
AEGL	acute exposure guideline level
ATSDR	Agency for Toxic Substances and Disease Registry
BW	body weight
CCEG	chemical casualty estimating guideline
CEGL	continuous exposure guidance level
CSF	cancer slope factors
CWA	chemical warfare agent
EEGL	emergency exposure guidance level
ERPG	emergency response planning guideline
FDWS	field drinking water standards
GSD	geometric standard deviation
HA	health advisory
HEAST	health effects assessment summary tables
HI	hazard index
HQ	hazard quotient
HSDB	hazardous substance databank
IOM	Institute of Medicine
IRIS	Integrated Risk Information System
LOAEL	lowest-observed-adverse-effect level
MAF	military adjustment factor
MCGL	maximum contaminant level goal
MCL	maximum contaminant level

MEG	military exposure guideline
MCRC	military cancer risk concentration
MF	modifying factor
MRC	military risk concentration
MRL	minimal risk level
NAAQS	national ambient air quality standards
NIOSH	National Institute of Safety and Occupational Health
NOAEL	no-observed-adverse-effect level
NRC	National Research Council
OEH/ED	occupational and environmental health/endemic disease
ORM	operational risk management
OSHA	Occupational Safety and Health Administration
PEF	particulate emission factor
PEGL	permissible exposure guideline level
PEL	permissible exposure limit
PMEG-L	preliminary military exposure guidelines–long-term
PPE	personal protective equipment
PRG	preliminary remediation goal
PSI	pollution standard index
RBC	risk-based concentration
REL	recommended exposure level
RfC	reference concentration
RfD	reference dose
SPEGL	short-term public guidance level
SSL	soil screening level
STEL	short-term exposure level
TEEL	temporary emergency exposure limit
THQ	target hazard quotient
TLV	Threshold Limit Value
UF	uncertainty factor
VOC	volatile organic compound

Review of the Army's Technical Guides on Assessing and Managing Chemical Hazards to Deployed Personnel

Summary

Military deployments involve a spectrum of activities that range from peacekeeping to combat. They are defined as troop movements to a land-based location outside of the continental United States that result from a Joint Chiefs of Staff/Unified Command deployment order applied for 30 or more consecutive days. In the past, deployment risk-assessment and risk-management strategies focused primarily on combat scenarios and warfare-related mission and health impacts. However, the roles of U.S. military forces have evolved and expanded. Increasingly, U.S. troops are deployed for operations other than war, including a variety of peacekeeping, humanitarian, and nation-building missions. Thus, the U.S. Department of Defense (DOD) now promotes a unified and comprehensive force health protection plan that advocates full consideration of all potential health hazards across the deployment spectrum and throughout the deployment process.

As part of mission planning, it is necessary for operational decision makers to have information on health hazards to individual soldiers and their potential impact on the options being considered for achieving the mission (i.e., the impacts on courses of action). The U.S. Army Center for Health Promotion and Preventive Medicine (USACHPPM) has developed guidance documents for assessing environmental health hazards that could be encountered during deployment. Technical Guide 248 (TG-248) provides a general approach to assessing chemical, radiological, physical, and endemic disease hazards, and Technical Guide 230 (TG-230) provides specific guidance on the chemical subset of hazards. The critical component of TG-230 is the use of military exposure guidelines (MEGs). MEGs are media- and duration-specific exposure values that indicate chemical concentrations in air, water, and soil at which certain adverse health effects might begin to occur in an exposed population. Documentation of how MEGs were derived for specific chemicals is provided in Reference Document 230 (RD-230).

STATEMENT OF TASK

The National Research Council (NRC) was asked to independently review TG-248, TG-230, and RD-230 for their scientific validity, completeness, and conformance to current risk-assessment practices. The NRC assigned this task to the standing Committee on Toxicology and convened the Subcommittee on Assessing Toxicological Risks to Deployed Military Personnel. The subcommittee was asked to review the Army's documents and to identify deficiencies and make recommendations for improvements. The subcommittee was asked to focus specifically on the following issues:

1. The Army's risk assessment, hazard-ranking, and risk-management processes described in TG-230 and its supporting documents.
2. The use of pre-existing exposure guidelines developed by the NRC and other agencies and organizations and the hierarchical scheme used by the Army in selecting from those various guidelines.
3. The Army's approaches to deriving MEGs for criteria pollutants, lead, soil contaminants, and other chemical contaminants.
4. Technical aspects of the Army's risk-management framework (as presented in TG-248) regarding competing health risks from different chemicals.
5. The assumption that the military population includes susceptible subpopulations (e.g., personnel with unknown health conditions, asthma, undetected pregnancies in the first trimester) and the use of uncertainty factors in the derivation of MEGs.
6. The adjustments of exposure guideline values to account for differences in exposure durations in the derivation of MEGs.
7. The exposure assumptions and mathematic models used for the derivation of MEGs for air, water, and soil contaminants.
8. Technical aspects of the Army's acceptable cancer risk level of 1 in 10,000.
9. The balance of emphasis between health effects that are produced immediately or soon after exposure and possible delayed effects (e.g., cancer) in the derivation of MEGs for chemical warfare agents and toxic industrial chemicals.

10. The use of a single risk-assessment methodology for assessing the toxicological risk from exposures to chemical warfare agents and toxic industrial chemicals rather than separate risk-assessment methodologies.

11. The assumption that the toxicity of a mixture of chemicals that have similar modes of action will be equal to the sum of the toxicities of individual chemicals in the mixture.

12. The utility of TG-248, TG-230, and RD-230 for decision makers (who might not be knowledgeable about toxicology or the science behind the health risk-assessment process) who will be using MEGs in the field.

THE ARMY'S PROCESS TO EVALUATE CHEMICAL HAZARDS

The goals of TG-230 are to "characterize the level of *health* and *mission* risks associated with identified or anticipated exposures to chemicals in the deployment environment" (italics added) so that chemical threats can be appropriately considered in operational planning. To achieve those goals, USACHPPM incorporated a risk-assessment matrix (see Table S-1)—a standard component of military operational risk management that is used for risk categorization—in its technical guides. This matrix is a qualitative classification tool that reflects four categories of severity in risk to a military mission and five categories of probability with regard to one or more military assets. The categories are used to characterize risk in terms of mission success.

The risk-assessment matrix was incorporated into TG-248 and TG-230 to facilitate the characterization of chemical hazards on the same basis and in the same terminology as other operational hazards (e.g., climate conditions, terrain, enemy forces). TG-230 uses MEGs as the basis for classifying the chemical hazards. MEGs are estimated concentrations of hazardous chemicals in air, water, or soil above which individuals might experience certain types of health effects after an exposure of specified duration. Measured or predicted concentrations of chemicals at the mission site are compared with the most relevant MEGs to determine the potential risks.

FINDINGS

Risk-Assessment, Hazard-Ranking, and Risk-Management Approaches

The framework developed by USACHPPM in TG-248 and TG-230 for

TABLE S-1 Risk-Assessment Matrix

		Probability				
Severity		Frequent A	Likely B	Occasional C	Seldom D	Unlikely E
Catastrophic	I	E	E	H	H	M
Critical	II	E	H	H	M	L
Marginal	III	H	M	M	L	L
Negligible	IV	M	L	L	L	L

Definitions

Hazard Severity

Catastrophic (I): Loss of ability to accomplish the mission or mission failure. Death or permanent disability. Loss of major or mission-critical system or equipment. Major property (facility) damage. Severe environmental damage. Mission-critical security failure. Unacceptable collateral damage.
Critical (II): Significantly degraded mission capability, unit readiness, or personal disability. Extensive damage to equipment or systems. Significant damage to property or the environment. Security failure. Significant collateral damage.
Marginal (III): Degraded mission capability or unit readiness. Minor damage to equipment or systems, property, or the environment. Injury or illness of personnel.
Negligible (IV): Little or no adverse impact on mission capability. First aid or minor medical treatment. Slight equipment or system damage, but fully functional and serviceable. Little or no property or environmental damage.

Risk Levels

E – Extremely high risk: Loss of ability to accomplish the mission if threats occur during mission. A frequent or likely probability of catastrophic loss (IA or IB) or frequent probability of critical loss (IIA) exists.
H – High risk: Significant degradation of mission capabilities in terms of the required mission standard, inability to accomplish all parts of the mission, or inability to complete the mission to standard if threats occur during the mission. Occasional to seldom probability of catastrophic loss (IC or ID) exists. A likely to occasional probability exists of a critical loss (IIB or IIC) occurring. Frequent probability of marginal losses (IIIA) exists.
M – Moderate risk: Expected degraded mission capabilities in terms of the required mission standard will have a reduced mission capability if threats occur during mission. An unlikely probability of catastrophic loss (IE) exists. The probability of a critical loss is seldom (IID). Marginal losses occur with a likely or occasional probability (IIIB or IIIC). A frequent probability of negligible (IVA) losses exists.
L – Low risk: Expected losses have little or no impact on accomplishing the mission. The probability of critical loss is unlikely (IIE), while that of marginal loss is seldom (IIID) or unlikely (IIIE). The probability of a negligible loss is likely or less (IVB through IVE).

Hazard Probability

Frequent (A): Occurs very often, continuously experienced.
Likely (B): Occurs several times.
Occasional (C): Occurs sporadically.
Seldom (D): Remotely possible; could occur at some time.
Unlikely (E): Can assume will not occur, but not impossible.

Unit Status

Black: Unit requires reconstitution. Unit below 50% strength.
Red: Combat ineffective. Unit at 50-69% strength.
Amber: Mission capable, with minor deficiencies. Unit at 70-84% strength.
Green: Mission capable. Unit at 85% strength or better.

Source: TG-230 and U.S. Army Field Manual 3-100.12.

assessing health and mission risks during deployment clearly attempts to address the recommendations made by DOD and past reports of the NRC and Institute of Medicine for developing a process that would incorporate consideration of all potential health hazards into operational decision making more thoroughly than was done in the past. The incorporation of the risk-assessment matrix into the guidance is essential because it builds health risk assessment into a process that is routinely used and is well understood throughout the military establishment. However, the subcommittee found that USACHPPM's approach of using one set of chemical exposure guidelines (the MEGs) was inadequate for achieving the two goals of assessing mission risks and providing force health protection. Table S-2 shows that the parameters for achieving those goals are different, which makes it extremely difficult for one set of guidance values to address both goals adequately.

MEGs were determined from pre-existing exposure guidelines designed to provide a reasonable assurance of safety by considering a diverse set of protective assumptions and addressing uncertainties conservatively. Thus, MEGs are appropriate (with some modification) for the goal of providing force health protection. However, for the assessment of chemical risks to missions, the goal is to provide an estimate of unit status (e.g., mission capable, combat ineffective, unit requires reconstitution) in the event of an exposure. Assessment of those hazards requires an understanding of casualty estimates—when soldiers' health and performance might be degraded to the extent that the mission is jeopardized. MEGs are inappropriate for making this type of assessment because they are estimates of concentration thresholds below which no adverse health effects are expected to occur, not estimates of concentrations at which mission-relevant casualties would occur. Thus, the MEG threshold concentrations are lower (perhaps even several orders of magnitude lower) than those at which mission-relevant casualties would be expected. For that reason, mission risk levels characterized on the basis of MEGs are not comparable to the risk levels assigned to other kinds of military operational hazards and could lead to overestimating the risk that chemicals pose to the mission. A second set of chemical exposure guidelines for mission-relevant casualty prediction is needed for the assessment of mission risks. How USACHPPM might develop another set of exposure guidelines to use in parallel with the MEGs is discussed later in this summary.

Use of Pre-Existing Exposure Guidelines

MEGs were developed by USACHPPM for contaminants in air, water, and soil. They were derived by reviewing guidelines and health-based criteria or

TABLE S-2 Characteristics Associated with the Major Goals of TG-248 and TG-230

	Health Risk Assessment	Mission Risk Assessment
Goal	To assess impacts on individual soldier health; requires the use of *protective* exposure values	To predict impacts of health risks on the mission; requires the use of *predictive* casualty estimates
Effects	Short- and long-term effects	Primarily short-term effects
Length of exposure	Long-term exposure	Short-term exposure
Situation	More like occupational/environmental (OSHA, EPA)	More like short-term emergency planning
Availability of data	More likely to have data available to assess exposure	More qualitative assessment of exposure; relies more on subjective judgment
Availability of time	More time to assess	Decisions must be made quickly
Exposure assessment	Assess proportion likely to receive exposure in excess of MEGs	Assess proportion likely to receive any mission-compromising level of exposure
Number of chemicals	Many of concern	Limited number of concern
Likelihood that effect(s) will occur	Lower	Higher
Confidence in estimated exposure(s)	Higher	Lower

standards from other agencies (e.g., the U.S. Environmental Protection Agency [EPA], the American Conference of Governmental Industrial Hygienists [ACGIH]), selecting the most relevant guidelines on the basis of a hierarchical scheme, and modifying the chosen guidelines for military use. The drawback of this approach is that the existing guidelines were designed to protect various populations that differ from deployed troops (e.g., the general population, workers) and were intended for different settings (e.g., ambient exposures, workplace, accidental releases), which made it necessary for USACHPPM to adjust the values to make them relevant to the deployment setting. The subcommittee found the application of these adjustments was not sufficient to ensure that the resulting values provide comparable levels of protection among various chemicals. In addition, the scientific basis of the MEGs was dependent on the data and methodology that were available when the existing guidelines were developed. These limitations are illustrated in the evaluation of the three chemical categories the Army specified for particular consideration—criteria air pollutants, lead, and soil contaminants.

- **Criteria air pollutants.** Criteria air pollutants were an important consideration for USACHPPM because they are ubiquitous and capable of causing adverse health effects in certain individuals at high ambient levels. In evaluating the long-term MEGs for those pollutants, the subcommittee found that the rationales for selecting one organization's guideline over another were questionable. In addition, adjustments intended to make the guideline relevant for military purposes were not applied consistently. For example, EPA's national ambient air quality standards (NAAQS) were used to derive a long-term MEG for carbon monoxide. The NAAQS for carbon monoxide was set to protect angina patients who exercise. No adjustments were made to account for the lack of such patients among the deployed military population. In another case, an occupational standard for sulfur dioxide was used and adjusted to derive a MEG. The resulting value was lower than the NAAQS for sulfur dioxide, which were designed by EPA to protect children and individuals with pre-existing lung disease. Establishing a MEG lower than the NAAQS requires some justification.
- **Lead.** For lead, the subcommittee discovered an error in the drinking water MEG—the World Health Organization's drinking water criterion, used as the basis for the MEG, was reported incorrectly by USACHPPM as 0.05 mg/L instead of 0.01 mg/L. The derivation of the soil MEG for lead also requires reanalysis, because the selected target blood lead concentration is not protective of the embryo and fetus. In contrast, many exposure assumptions used in the adult blood level model are overly conservative.
- **Soil contaminants.** For soil contaminants, the subcommittee identified a number of concerns, including the use of older data or assumptions that have been recently updated or have been superceded by new guidance; failure to develop MEGs for volatile chemicals; flaws in the description of how dermal toxicity values are derived; and the use of a high and uncertain soil ingestion rate. In addition, soil MEGs were only established for 1-year exposures, and the subcommittee is not convinced that short-term soil MEGs are unnecessary. There are certain chemicals, such as volatile chlorinated solvents, for which short-term soil MEGs would be appropriate to protect troops in trenches or in tents above contaminated soil.

Adjustments for Military Application

In developing TG-230 and MEGs, it was necessary for USACHPPM to make assumptions about the composition of its forces, to factor in deployment exposure conditions, and to select an acceptable lifetime cancer risk.

Population Considerations

USACHPPM considered demographic and health differences between deployed populations and the general public that might contribute to differences in susceptibilities to environmental exposures. For the general public, susceptible subpopulations typically include embryos and fetuses, the very young, the elderly, individuals with pre-existing disease, and those with genetic susceptibilities. In establishing health-protective exposure values, uncertainty factors are conventionally applied to provide a margin of safety to protect the portion of the general population that might be at increased risk. In some cases, such as the NAAQS noted above, values were based on data from a susceptible subpopulation.

According to the demographic information provided to the subcommittee, with the exception of genetic susceptibilities, deployed military personnel include few individuals in the traditional categories of increased susceptibility relative to the general population. Deployed personnel span a narrower age range and are subject to physical requirements that should ensure that they are in better health or do not have pre-existing medical conditions that might interfere with their ability to serve during a deployment. Although TG-230 identifies asthmatic individuals as a subgroup that might be more susceptible to certain contaminants, it appears to the subcommittee that documentation and procedures are in place that would prevent or limit the deployment of asthmatic personnel, especially those with moderate or severe disease. Thus, the subcommittee concludes that the deployed forces should be considered healthier than the general public. On the other hand, it is reasonable to assume that the deployed military population might have a level of genetic susceptibilities similar to that found in the general population.

TG-230 indicates that although women known to be pregnant are excluded from deployment, there could be cases where a pregnancy is discovered only after deployment. In such situations, it is possible for exposures to occur during critical stages of embryo and fetal development before pregnant women have been removed from the deployment scenario. In addition, some chemicals could persist in the body after deployment and have a potential to affect post-deployment pregnancies. Thus, it is important that MEGs be protective against developmental effects. However, it was unclear to the subcommittee whether all of the chemicals had been screened for developmental effects. For example, some of the extant military exposure guidelines (such as the military's continuous exposure guidance levels) are set on the basis of an assumption that only men would be exposed, so developmental effects were not considered.

Exposure Adjustments

USACHPPM had to adjust existing guidelines set by regulatory and other agencies for application in the military context. Key adjustments were made for exposure rates and differences in military population characteristics compared with the general population. For example, the activity level of deployed troops is much higher than that of the general population, such that breathing and water-consumption rates of military personnel are much higher. To account for exposure differences, simple mathematic adjustments were used. The subcommittee found that USACHPPM provided adequate justification for performing those exposure adjustments but appears to have applied them inconsistently in some cases. For example, the inhalation adjustment factor appears to have been used in setting some, but not all, of the 14-day air MEGs.

Cancer Risk

The Army posed the question of whether a cancer risk of 1 in 10,000 is acceptable for establishing MEGs for carcinogens. The identification of an acceptable cancer risk level has been debated for many years. It is essentially a risk-management policy decision, because the selection of an acceptable risk is a question of societal norms and values. Consequently, science does not directly provide an answer to the question. The subcommittee concluded that it would be inappropriate for it to make a judgment about how much risk the military should accept. However, the subcommittee decided it could help address that issue by reviewing acceptable risk levels selected by other organizations and making observations about where the Army's acceptable cancer risk threshold lies in comparison. The subcommittee found that risk of 1 in 10,000 falls within the range used by other federal agencies for occupational and environmental exposures, and is sufficiently conservative to be protective for individual soldiers in the event of multiple deployments.

Immediate and Delayed Health Effects

The subcommittee was asked to evaluate whether appropriate consideration was given to immediate and delayed health effects. It was clear that USACHPPM considered long-term health consequences along with short-term effects during the development of the chemical hazard ranking scheme for mission risk assessment. However, the subcommittee found that a more formalized procedure for communicating long-term and delayed health effects simultaneously with mission risk information is needed to ensure that those potential effects are explicitly considered. In addition, the discussion of delayed effects highlights cancer, and places inadequate emphasis on other chronic or delayed effects (e.g., compromised immune function, infertility).

Use of a Common Risk-Assessment Methodology

One of the questions posed to the subcommittee was whether chemical warfare agents should be evaluated differently from toxic industrial chemicals. The subcommittee found no reason not to apply the same risk-assessment methodology to those two categories of chemicals. Chemicals will have to be evaluated on a case-by-case basis, but the risk-assessment approach to evaluating them can and should be conceptually similar.

Exposure Assessment

MEGs are designed to be compared with measured or modeled concentrations in the field. The subcommittee was informed that intelligence information on potential sources of chemical hazards is generally available for making predeployment risk assessments and that procedures are in place for conducting environmental sampling during deployments. However, no references are provided in TG-248 or TG-230 to documentation on how such information is to be collected and assessed. It seems appropriate that risk analysts and preventive-medicine personnel would be involved, to some extent, in developing exposure-assessment plans; therefore, it would be helpful if exposure-assessment guidance was compiled from existing sources and incorporated into or at least linked with TG-230 to support those personnel. The guidance should include information on exposure monitoring and modeling and on developing a sampling plan.

Cumulative Risk

Cumulative risk is the likelihood of occurrence of an adverse health effect resulting from exposure to multiple chemicals that have common modes of toxicity from all routes and pathways. Assessing cumulative risk is a complex task that requires assessing whether the toxic effects of chemicals found in a mixture produce their effects independently or produce additive, synergistic, or antagonistic effects. TG-230 assumes that total toxicity from chemicals in a mixture of toxicants with similar modes of action is equal to the sum of the weighted dose toxicities of the individual chemicals. Although that generally is accepted practice, no guidance was provided on how the cumulative risks were to be assessed other than by a qualitative notation. The subcommittee examined a number of chemicals in TG-230 that have effects that would at least summate with one another, but found that it was impossible to identify that type of potential additive action from the descriptions of symptoms and target organs provided in RD-230. As a first step toward improving the assessment of cumulative risks, it might be practical to establish a qualitative classification scheme that identifies chemicals known to interact or cause similar effects and that might be encountered simultaneously during a deployment. Then USACHPPM can consider incorporating quantitative approaches.

Utility for Decision Makers

The subcommittee was asked to consider whether the technical guides could be used by personnel who are not knowledgeable about toxicology or risk assessment. Although the technical guides provide a procedure that is intended to facilitate the consistent evaluation and interpretation of chemical threats that might be encountered during deployment, the subcommittee found that professional judgment of trained personnel is necessary to use the guides properly and to effectively communicate risks to nontechnical decision makers. Another element of the task question is whether the products of TG-230 will be understood by decision makers so they can properly consider risks to the mission and to force health. As noted earlier, the current set of exposure guidance does not allow for adequate characterization of mission risks and will, therefore, lead to decisions based on inappropriate comparisons between chemical risks and other operational risks.

RECOMMENDATIONS

Use Two Sets of Exposure Guidelines

The subcommittee recommends that two sets of chemical exposure values be used to assess health risks and mission risks separately. This will ensure that the guideline values are based on health considerations appropriate to the intended goal. Below, the subcommittee outlines how the two sets of guidelines should be derived and applied in the operational risk-management process.

Exposure Guidelines for Assessing Mission Risks

One goal of TG-248 and TG-230 is to characterize the levels of mission risk posed by chemicals for comparison with other operational risks. To address this goal, the subcommittee recommends the Army develop a new set of chemical exposure guidelines that provide predictive estimates of mission-relevant casualties in the event of an exposure during a mission. Such values, termed chemical casualty estimating guidelines (CCEGs) by the subcommittee, would be defined as media- and duration-specific chemical concentrations expected to cause health impairments that degrade the performance of enough individuals to reduce unit strength. CCEGs should not be established from existing health-protective exposure standards, but should be derived by conducting independent evaluations of each chemical of interest and developing exposure-response and population-response data on which to base casualty estimations. Using casualty estimates (rather than health protective estimates) in conjunction with the operational risk-assessment matrix will provide risk-level characterizations more appropriate for comparison with other anticipated risks as well as with other chemical hazards. The following are important elements to consider in developing CCEGs:

- A methodology should be developed to derive CCEGs that provide predictive, probabilistic exposure-response information that would enable decision makers to weigh chemical threats in comparison to other mission threats as well as to other chemical hazards. CCEGs ideally would be determined by modeling chemical-specific data to predict effects on unit strength at various exposure levels (e.g., probit analysis, which provides a graphic representation of a dose-response relationship in the ranges where effects are observed).

- CCEGs should be established for chemicals that have some finite probability of being encountered in sufficient quantities to degrade a mission.
- CCEGs should be derived primarily for air contaminants, because inhalation is the exposure route most likely to result in incapacitation. However, there are some situations for which oral and dermal CCEGs might be necessary, such as specialized operation activities that involve exposure to contaminated water (e.g., water immersion activities).
- Assistance should be solicited from other agencies and organizations working on health-related guidelines. Many existing exposure guidelines (especially EPA's acute exposure guideline levels) make key information readily available. Future working relationships between the DOD and other agencies routinely developing exposure guidelines might make the development of CCEGs more resource-effective.
- The methodology for deriving CCEGs and the derivation of the CCEGs themselves should be peer-reviewed.
- If the Army chooses to use MEGs in the interim, TG-230 should be revised to warn users regarding the deficiencies and limitations of MEGs when applied to assess mission-related performance risks.

Exposure Guidelines for Assessing Health Risks

Another goal of TG-248 and TG-230 is to provide force health protection across a range of scenarios that might be encountered during deployment, recognizing that some health risks might have to be accepted to achieve military objectives. The subcommittee found that the MEGs are conceptually appropriate for addressing health threats in terms of force health protection. However, the procedures for developing MEGs outlined in RD-230 require some modification to make the MEGs more relevant to deployment situations and more consistently protective. In addition, guidance should be added to TG-230 on how to apply and interpret the MEGs. The following are important elements to consider in addressing this recommendation:

- Ideally, USACHPPM would benefit from developing an independent set of principles and procedures to develop MEGs from the available toxicology data on individual chemicals. Those procedures would solidify the purpose of the MEGs and would make explicit the risk-management policy decisions that underlie the selection of studies and use of uncertainty factors that might be different from those used by other agencies. However, the subcommittee recognizes the immensity of such an undertaking and therefore suggests that revisions be conducted in a prioritized manner. Below is a general description of the types of revisions needed. Specific examples and recommendations are provided in Chapter 5 of the report.

  —**Near-term revisions**. These are revisions to improve the quality of the MEGs that require relatively modest resources. They include revising the MEGs with updated values from other organizations, ensuring consistent use of uncertainty factors and adjustments relevant to the deployed population, ensuring that the MEGs are not based on data from subpopulations not expected to be among the deployed forces (e.g., asthmatics, children), and improving the documentation and use of the most relevant toxicity end points and uncertainty factors in setting the existing exposure guidelines.

  —**Mid-term revisions**. Revisions in this category would result in more internally consistent MEGs that are relevant to deployed populations. Such revisions would involve using original source material (e.g., the critical paper selected by EPA for a reference concentration) to calculate MEGs. MEGs should also be reviewed to assess whether they protect against developmental effects.

  —**Long-term revisions**. These include developing more rigorous procedures for determining MEGs and performing the analyses. The potential for collaboration with other agencies that are developing exposure guidelines should be explored. For example, EPA is beginning a major effort to update its Integrated Risk Information System. It might be possible to work with other agencies to establish deployment-relevant guidelines.

- USACHPPM should develop a risk-management framework that focuses on what action plans (i.e., responses) are appropriate when MEGs are exceeded. Possible responses would include considering risk-management options for reducing or eliminating risks (e.g., using protective gear, moving deployed personnel to an uncontaminated area, treating water) and determining the appropriate medical follow-up (e.g., documenting the exposure in medical records, tracking exposed individuals, providing long-term care) when military personnel must bear health risks.

Communication of Mission and Health Risks

Because some of the decisions that must be made with the guidance of TG-230 require subjective evaluation, it is important that personnel using the guides include individuals with training in preventive medicine, toxicology, and risk assessment. Trained personnel should conduct separate analyses of health and mission risks. The resulting evaluations should be provided to decision makers simultaneously and consideration should be given to the risk-management options available for reducing or eliminating the risks. This will help decision makers explicitly balance the competing health and mission risks with respect to the military objective. It also will help to ensure that any risk trade-offs that involve accepting some health risks to deployed personnel are recognized and that appropriate medical attention, surveillance, and follow-up are provided.

1

Introduction

BACKGROUND

Deploying military personnel in hostile or unfamiliar environments is inherently risky. Unlike garrison environments, which are reasonably well-protected, well-known, and well-controlled, deployment environments are imposed by the military mission. Deployment can present a novel array of military and nonmilitary threats, and mission objectives often dictate that those threats be addressed. Many deployment activities are not routine. Tasks must be accomplished with limited means, despite the potential dangers of the setting. In the deployment environment, time, materiel, and attention are at a premium, and excessive precautions can engender their own risks or jeopardize the military mission.

In the past, health-based risk-assessment and risk-management strategies for deployment situations focused primarily on warfare-related mission impacts. However, recent wars and conflicts, such as operations Desert Shield and Desert Storm, have highlighted the need for the U.S. military to protect its forces from health threats that do not directly impact the mission, are indirectly related to battle, or could appear after the deployment has ended. Thus, the U.S. Department of Defense (DOD) has developed a force health protection plan that is a "unified and comprehensive strategy that aggressively promotes a healthy and fit force and provides full protection from all potential health hazards throughout the deployment process. Its major ingredients include healthy and fit force promotion, casualty and injury prevention, and casualty care and management" (U.S. Department of the Army 2001). "Deployment" is defined as a troop movement to a land-based location outside the continental United States that does not have a permanent medical treatment facility (i.e., funded by the Defense Health Program). Deployment is the result of a Joint Chiefs of Staff/Unified Command deployment order and lasts for 30 or more consecutive days (U.S. Department of the Army 2001).

The role of U.S. military forces has changed and expanded. Increasingly, U.S. troops are deployed for operations other than war, including peacekeeping, humanitarian, and nation-building missions of varying scope and duration. (See Figure 1-1 for an illustration of potential conflicts and likelihood of occurrence.) Deployments differ in the degree and nature of tactical risks (i.e., risk due to the presence of an enemy or adversary). However, with or without tactical threats, there are risks of accident, disease, and illness inherent in deployment. Those might arise from contaminated local environments, from the intensive activities of the deployed forces, from exposure to hazards associated with mission tasks, from intentional exposures to pesticides and prophylactic agents, and from the rigors of exposure to climatic extremes.

In deployment situations, commanders must balance the effects of multiple risks. Effects can include casualties, impacts on civilians, damage to the environment, loss of equipment, and levels of public reaction against the value of the mission objectives. The Army's Field Manual 100-14 (U.S. Department of the Army 1998) outlines the principles, procedures, and responsibilities of applying an operational risk-management (ORM) process to conserve combat power and resources. The manual defines risk management as "the process of identifying, assessing, and controlling risks arising from operational factors and making decisions that balance risk costs with mission benefits . . . . It applies to all missions and environments across the wide range of Army operations." The ORM process is a cycle of (1) identifying hazards, (2) assessing the risk associated with those hazards, (3) developing controls and making risk decisions, (4) implementing the controls, and (5) supervising and evaluating the effectiveness of the controls. The process is depicted in Figure 1-2. The basic principles for implementing the process include the following:

- **Integrating risk management into mission planning, preparation, and execution.** Leaders and staff continuously identify hazards and assess both accidental and tactical risks. They then develop and coordinate control measures. They determine the level of residual risk for accidental hazards in order to evaluate courses of action, and they integrate control

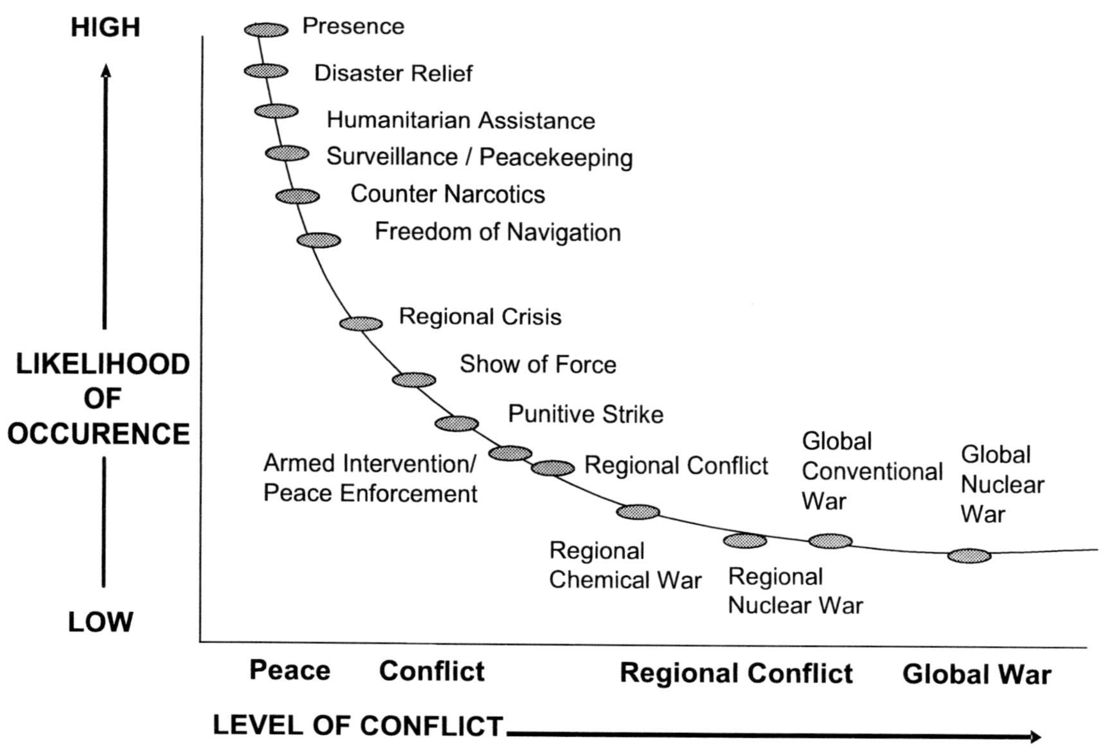

FIGURE 1-1 Spectrum of potential conflict. Source: Ciesla 2002.

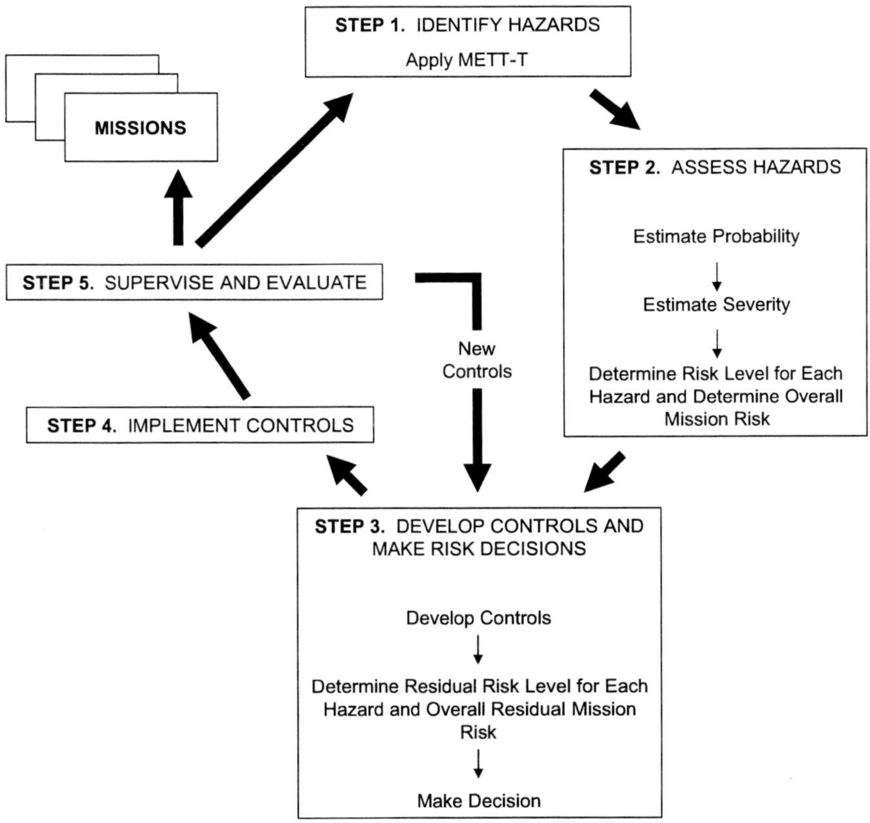

FIGURE 1-2 Continuous application of risk management. Source: Modified from U.S. Department of the Army 1998.

measures into staff estimates, operational plans (OPLANs), operation orders (OPORDs), and missions. Commanders assess the areas in which they might take tactical risks. They approve control measures that will reduce risks. Leaders ensure that all soldiers understand and properly execute risk controls. They continuously assess variable hazards and implement new risk controls.

- **Making risk decisions at the appropriate level in the chain of command.** The commander should address risks in his guidance. He should base his risk guidance on established Army and other appropriate policies and on his higher commander's direction. He then gives guidance on how much risk he is willing to accept and delegate. Subordinates seek the commander's approval to accept risks that might imperil the next higher commander's intent.

- **Accepting no unnecessary risk.** Commanders compare and balance risks against mission expectations and accept risks only if the benefits outweigh the potential costs or losses. Commanders alone decide whether to accept the residual risk to accomplish the mission.

As part of the DOD's force health protection program, the U.S. Army is developing strategies and methods for assessing the broad range of potential occupational and environmental health (OEH) threats that might occur as a result of deployment. Those threats include chemical, radiological, biological, entomological, and endemic-disease hazards. In the past, Army policies addressed health threats under only two deployment conditions—garrison peacetime deployment and wartime deployment. No guidance was available for the range of deployments that fall between those mission extremes. Recognizing this need for guidance, the Army developed an OEH policy intended to address the broad spectrum of possible military operations, activities, and scenarios (U.S. Department of the Army 2001). The goal of the policy is to allow commanders to make informed decisions about OEH hazards and there by minimize the total risk to soldiers and civilian personnel executing a range of military operations. To help commanders consider chemical OEH threats in their strategic decision-making process, the Army has developed two technical guides (Technical Guide 230 and Technical Guide 248) and one reference guide (Reference Document 230) that propose methods for assessing and managing chemical risks to deployed personnel. This NRC report reviews those documents for their scientific validity and their conformance with current understanding of risk-assessment practices.

The technical guides and reference document were informed by the efforts of several task forces and committees that have spent years evaluating the health of veterans; ensuring appropriate evaluation and care of veterans' health concerns; determining connections between service in the Persian Gulf, specific exposures, and veterans' health status; and developing guidance to help prevent and reduce unanticipated illnesses in future deployments. Reports from these groups include Presidential Review Directive 5 (NSTC 1998), *Potential Radiation Exposure in Military Operations* (IOM 1999a), *DOD Strategy to Address Low-Level Exposures to Chemical Warfare Agents* (DOD 1999), and *Protecting Those Who Serve* (IOM 2000) and its supporting reports, *Strategies to Protect the Health of Deployed U.S.*

Forces (IOM 1999b; NRC 1999; 2000a,b). The following considerations from those reports were instrumental in shaping the Army guidance:

- Full range of deployment scenarios (as illustrated in Figure 1-1).
- Broad range of chemical types, including those that are unique to the military (e.g., chemical warfare agents, depleted uranium, smokes, and obscurants) and those that might be present at the deployment site (e.g., pesticides, toxic industrial chemicals).
- Low-level exposures. The military has focused on exposures to high concentrations of chemicals because those exposures are the most likely to have direct negative consequences on the success of missions. However, more attention is being given to possible health effects from exposures to low concentrations of chemicals, particularly exposures that occur over an extended period of time.
- Personnel assumptions. In the past, deployed military populations were assumed to consist of healthy, physically fit men and nonpregnant women. Although personnel must meet certain health and fitness requirements, the military now recognizes that deployed populations (active duty, reserve, and National Guard personnel) can include individuals with health factors that might make them more susceptible to certain chemicals.
- Broad range of health effects. Historically, the military primarily was concerned with health threats that would affect deployed personnel immediately, because those might have the potential to affect the success of the mission. Operational planning now includes more emphasis on considering the risk of health effects that could occur months or even years after exposure.

GUIDANCE DOCUMENTS

Technical Guide 248 (TG-248) (USACHPPM 2001) proposes processes and tools to be used by preventive-medicine personnel for evaluating and communicating the occupational and environmental health (OEH) and endemic disease (ED) risks of deployment to commanders in accordance with the Army's ORM process. The process it proposes is intended to (1) document OEH/ED hazards and exposures to soldiers and the force, (2) characterize the risk of OEH/ED hazards during all phases of deployment, (3) communicate risks in understandable terms to commanders and operational planners, (4) allow the commanders' staffs to develop courses of action that consider and/or minimize OEH/ ED risks to the force, and (5) provide data to assist in post-deployment health assessments and evaluations of OEH/ED operational risk-management processes. The overall goal of TG-248 is to characterize OEH/ED risks in such a way that they can be placed in a similar ranking scale with each other and with other operational hazards. This report focuses on the usefulness of TG-248 for evaluating *chemical hazards only,* and not radiological, biological, entomological, or endemic disease hazards.

Technical Guide 230 (TG-230) (USACHPPM 2002a) presents proposed military exposure guidelines (MEGs) for chemicals in air, water, and soil. A MEG is an estimated chemical concentration above which certain types of adverse health effects might begin to occur in individuals within the exposed population after a continuous, single exposure of specified duration. MEGs are used for deployment purposes only and are different from occupational standards for garrison situations. MEGs are used to assess the significance of field exposures to chemical hazards during deployment. They are designed to address a variety of exposure conditions not covered by occupational or other standards used in garrisons, such as a single catastrophic release of large amounts of a chemical, temporary exposures lasting hours or days, continuous ambient environmental conditions (e.g., regional pollution), use of a contaminated water supply, or persistent soil contamination.

MEGs were developed for chemicals for which information was readily available and for chemicals that were otherwise identified by the Army as key hazards of concern, including chemical warfare agents and toxic industrial chemicals. For air contaminants, the Army developed MEGs for exposure durations of 1 hour, 8 hours, 24 hours, 14 days, and 1 year. For water contaminants, MEGs were developed for exposure durations of 5 days, 14 days, and 1 year. For soil contaminants, only 1-year MEGs were developed, because short-term exposure guidelines were deemed unnecessary. The Army does not anticipate that soil contamination will be an immediate or severe hazard. Severely contaminated soils are often easily detected because of odors, dead or discolored vegetation, or free chemical product.

TG-230 proposes a standardized process for using MEGs to characterize the levels of health and mission risk associated with chemical exposures in accordance with the military's ORM paradigm. The guidance is intended for use by preventive-medicine personnel, environmental staff officers, industrial hygienists, health risk assessors, and other medically trained personnel. An important element of the assessment process outlined in TG-230 involves the distinction between a "health threat" and a "medical threat." A health threat would affect an individual soldier's health, whereas a medical threat refers to a subset of health threats that have the potential to degrade a unit's combat (or mission) effectiveness.

TG-230 is supported by Reference Document 230 (RD-230) (USACHPPM 2002b), which provides details of the scientific rationale and assumptions that were used to derive the MEGs. The general approach for deriving MEGs was to select the most relevant existing exposure guidelines or peer-reviewed toxicological estimates developed for workers and the general population by government agencies or other organizations and to accept or adjust those values for deployment scenarios. The Army selected that approach because it was the most expedient and least costly way to develop exposure guidance for a large number of chemicals.

STATEMENT OF TASK

The National Research Council (NRC) was asked to independently review TG-248, TG-230, and RD-230 for their scientific validity, completeness, and conformance to current risk-assessment practices. The subcommittee was asked to review the Army's documents, identify deficiencies, and make recommendations for improvements. The subcommittee was asked to focus specifically on the following issues:

1. The Army's risk assessment, hazard-ranking, and risk-management processes described in TG-230 and its supporting documents.
2. The use of pre-existing exposure guidelines developed by the NRC and other agencies and organizations and the hierarchical scheme used by the Army in selecting from those various guidelines.
3. The Army's approaches to deriving MEGs for criteria pollutants, lead, soil contaminants, and other chemical contaminants.
4. Technical aspects of the Army's risk-management framework (as presented in TG-248) regarding competing health risks from different chemicals.
5. The assumption that the military population includes susceptible subpopulations (e.g., personnel with unknown health conditions, asthma, undetected pregnancies in the first trimester) and the use of uncertainty factors in the derivation of MEGs.
6. The adjustments of exposure guideline values to account for differences in exposure durations in the derivation of MEGs.

7. The exposure assumptions and mathematic models used for the derivation of MEGs for air, water, and soil contaminants.

8. Technical aspects of the Army's acceptable cancer risk level of 1 in 10,000.

9. The balance of emphasis between health effects that are produced immediately or soon after exposure and possible delayed effects (e.g., cancer) in the derivation of MEGs for chemical warfare agents and toxic industrial chemicals.

10. The use of a single risk-assessment methodology for assessing the toxicological risk from exposures to chemical warfare agents and toxic industrial chemicals rather than separate risk-assessment methodologies.

11. The assumption that the toxicity of a mixture of chemicals that have similar modes of action will be equal to the sum of the toxicities of individual chemicals in the mixture.

12. The utility of TG-248, TG-230, and RD-230 for decision makers (who might not be knowledgeable about toxicology or the science behind the health risk-assessment process) who will be using MEGs in the field.

THE SUBCOMMITTEE'S APPROACH

To accomplish its task, the subcommittee held four meetings between October 2002 and August 2003. The first two meetings involved data-gathering sessions that were open to the public. The subcommittee heard presentations from DOD, on its force health protection program, and from the U.S. Army Center for Health Promotion and Preventive Medicine, the service organization responsible developing TG-230, RD-230, and TG-248. The subcommittee critically evaluated TG-230, RD-230, and TG-248 as well as other supporting documentation from the Army. The documents were evaluated for their technical soundness, conformance with current risk-assessment practice, and utility for the intended user.

This report is organized into four chapters. Chapter 2 reviews the framework provided in TG-248 and TG-230 for assessing and managing mission and health risks from chemical exposures. Chapter 3 reviews the key concepts, assumptions, and decisions made in developing TG-248, TG-230, and RD-230. Chapter 4 outlines the subcommittee's recommended approach to characterizing mission risks, and Chapter 5 presents how MEGs should be improved to support health risk assessment and determine health risk management options. Table 1-1 presents a list of tasks and the corresponding chapters and relevant pages.

TABLE 1-1 Chapters That Address the Specific Task Issues

Task	Location
1. The Army's risk-assessment, hazard-ranking, and risk-management processes	Chapter 2
2. Use of existing exposure guidelines and the hierarchy for their selection	Chapter 2, general overview Chapter 5, medium-specific guidelines and hierarchies reviewed
3. Criteria pollutants, lead, and soil contaminants	Chapter 3, lead Chapter 5, criteria pollutants, lead, and soil
4. Competing health risks from other hazards and between chemicals	Chapter 2
5. Assumptions about the military population and the use of uncertainty factors	Chapter 3
6. Adjustments for exposure durations	Chapter 5, medium-specific adjustments reviewed
7. Exposure assumptions and calculations used to develop MEGs	Chapter 3, general overview Chapter 5, medium-specific assumptions and calculations reviewed
8. Acceptable cancer risk of 1 in 10,000	Chapter 3 Appendix B
9. Balance between immediate and delayed or chronic health effects	Chapter 3
10. Use of a common risk-assessment methodology for chemical warfare agents and toxic industrial chemicals	Chapter 3
11. Chemical mixtures	Chapters 3 and 4 Chapter 5, possible approaches for MEGs Appendix E, possible approaches for CCEGs
12. Utility for decision makers	Chapter 3

REFERENCES

Ciesla, J.J. 2002. Military Operational Deployments. An Information Brief for the National Research Council. Presentation at the First Meeting on Toxicological Risk to Deployed Military Personal, October 2, 2002, Washington, DC.

DOD (U.S. Department of Defense). 1999. DOD Strategy to Address Low-Level Exposures to Chemical Warfare Agents (CWAs). May 1999. [Online]. Available: http://chppm-www.apgea.army.mil/chemicalagent/caw/lowlevestrategy.PDF [accessed November 25, 2003]

IOM (Institute of Medicine). 1999a. Potential Radiation Exposure in Military Operations: Protecting the Soldier Before, During, and After. Washington, DC: National Academy Press.

IOM (Institute of Medicine). 1999b. Strategies to Protect the Health of Deployed U.S. Forces: Medical Surveillance, Record Keeping, and Risk Reduction. Washington, DC: National Academy Press.

IOM (Institute of Medicine). 2000. Protecting Those Who Serve: Strategies to Protect the Health of Deployed U.S. Forces. Washington, DC: National Academy Press.

NRC (National Research Council). 1999. Strategies to Protect the Health of Deployed U.S. Forces: Force Protection and Decontamination. Washington, DC: National Academy Press.

NRC (National Research Council). 2000a. Strategies to Protect the Health of Deployed U.S. Forces: Analytical Framework for Assessing Risks. Washington, DC: National Academy Press.

NRC (National Research Council). 2000b. Strategies to Protect the Health of Deployed U.S. Forces: Detecting, Characterizing, and Documenting Exposures. Washington, DC: National Academy Press.

NSTC (National Science and Technology Council). 1998. A National Obligation Planning for Health Preparedness for and Readjustment of the Military, Veterans, and Their Families after Future Deployments. Presidential Review Directive 5, Executive Office of the President, Office of Science and Technology Policy, Washington, DC.

USACHPPM (U.S. Army Center for Health Promotion and Preventive Medicine). 2001. Guide for Deployed Preventive Medicine Personnel on Health Risk Management. Technical Guide 248. U.S. Army Center for Health Promotion and Preventive Medicine. August 2001. [Online]. Available: http://chppm-www.apgea.army.mil/deployment/ [accessed November 25, 2003].

USACHPPM (U.S. Army Center for Health Promotion and Preventive Medicine). 2002a. Chemical Exposure Guidelines for Deployed Military Personnel. Technical Guide 230. U.S. Army Center for Health Promotion and Preventive Medicine. January 2002. [Online]. Available: http://chppm-www.apgea.army.mil/deployment/ [accessed November 25, 2003].

USACHPPM (U.S. Army Center for Health Promotion and Preventive Medicine). 2002b. Chemical Exposure Guidelines for Deployed Military Personnel. A Companion Document to USACHPPM Technical Guide (TG) 230 Chemical Exposure Guidelines for Deployed Military Personnel. Reference Document (RD) 230. U.S. Army Center for Health Promotion and Preventive Medicine January 2002. [Online]. Available: http://chppm-www.apgea.army.mil/deployment/ [accessed November 25, 2003].

U.S. Department of the Army. 1998. Risk Management, Field Manual No. 100-14. U.S. Department of the Army, Washington, DC. April 23, 1998.

U.S. Department of the Army. 2001. Force Health Protection (FHP): Occupational and Environmental Health (OEH) Threats. HQDA Ltr 1-0-1. U.S. Department of the Army, Washington, DC. June 27, 2001.

2

Review of the Army's Technical Guidance

This chapter reviews the approach used in Technical Guide 248 (TG-248) and Technical Guide 230 (TG-230) for assessing and characterizing chemical hazards for deployment decision making. First, important aspects of risk comparison that were identified in previous reports of the National Academies are briefly revisited. Second, the adequacy of the technical guides for providing appropriate characterizations of health and mission risks from exposure to chemicals is evaluated.

EARLIER ACADEMIES REPORTS ON DEVELOPING RELIABLE COMPARATIVE RISK ASSESSMENTS FOR DEPLOYMENTS

The Institute of Medicine report *Protecting Those Who Serve* (IOM 2000) and the National Research Council report *Strategies to Protect the Health of Deployed U.S. Forces: Analytical Framework for Assessing Risks* (NRC 2000) provide a number of recommendations relevant to developing a systematic process to prospectively evaluate non-battle-related risks associated with deployment activities and settings. For example, NRC (2000) specifically recommends that the conceptual paradigm for quantitative risk assessment described in *Risk Assessment for the Federal Government: Managing the Process* (NRC 1983) be used as a basis for developing a U.S.

Department of Defense (DOD) framework for assessing risks to deployed forces. Use of that paradigm would "facilitate integration of the results of hazard-specific assessments and tracking of the complex process of simultaneous consideration of multiple threats … and [would aid in] developing risk management strategies, including trade-offs" (NRC 2000). That recommendation follows from a more detailed discussion of related issues in which the following key points are made:

- "Troops during deployment could become exposed to a number of threats simultaneously. Exposures that are individually tolerable without appreciable risk might not be so when several are experienced together, and the question of interactions among agents looms particularly large for deployment risk assessment" (NRC 2000, p. 41)
- "The NRC (1983) paradigm for risk assessment … is readily adaptable to deployed forces protection … to analyze (1) the likelihood of the presence of a hazard associated with a deployment; (2) the likelihood of releases of agents into the environment; (3) the likelihood that troops will suffer exposure (of various magnitudes), given the releases; and (4) the likelihood that health effects will be caused among them, given the exposure. … [E]fforts would be focused on how activities and practices come to present threats, how likely it is that threats will be manifested in practice, and how mitigating one risk might raise other risks" (NRC 2000, p. 43)
- "[R]isk analysis must be content to say what can be said and not only to acknowledge the inevitable remaining uncertainty, but to try to characterize that uncertainty so that appropriate perspectives on the meaning and robustness of the analysis are expressed. … Characterization of uncertainty and the limitations of available data are important to all risk analysis, but they might play an especially important role in the analysis of deployment threats, where high-consequence decisions might require taking one risk to avoid others, Risk management approaches exist to help make such decisions, but *when the risks to be compared are quite uncertain, or uncertain to different degrees, good characterizations of uncertainty is [sic] necessary in order to arrive at sound solutions*" (NRC 2000, pp. 60-61; italics added).
- "… the establishment of 'conservative' estimates of dose-response relations, that is, those designed to err on the side of safety when faced with uncertainty about how to project expected human responses from available data, might not be appropriate for certain military uses. *When risks cannot be avoided and decisions are made to accept some risks rather than others*, or to bear some risk in furtherance of a more fundamental military objective, *it is important to make these trade-off decisions with unbiased esti-*

mates of the impacts of various courses of action. In other applications, such as the setting of health-protective exposure standards for application in less severe circumstances, conservative estimates might be much more acceptable. ... [Analyses should be] conducted and ... results presented, so that different uses appropriate for different risk-management settings can be made" (NRC 2000, pp. 66-67; italics added).

- "A final *special aspect of risk analysis for deployment is the large role that risk-risk comparisons must play.* Given the high level of tactical risk that might be inherent in the deployment situation, some health and safety risks may be appropriate to avoid or mitigate even greater risks. Determining how to optimize the trade-offs requires simultaneous consideration of the spectrum of risks faced by deployed troops, along with the possibility that actions taken to avoid or ameliorate some risks might exacerbate others" (NRC 2000, pp. 83-84; italics added).
- "[DOD decisions concerning deployed military personnel involve issues including] the need to call for individual troops to put life, limb, and health at risk in the interests of the military mission and the nation at large; [and] problems of trading off possibilities of health effects in later life with immediate risks of casualties and impacts on military mission of military capabilities ... *If the risk analysis is to effectively contribute to such decisions, it will require an articulation of a doctrine on how risk trade-offs are to be considered.* In addition, DOD should attempt to articulate a set of principles on how the balance of long-term risks to the troops and risks to the military mission should be approached" (NRC 2000, p. 89; italics added).

These key points highlight the importance of using a comprehensive, quantitative risk-assessment paradigm as the basis for a formal framework for the integrated management of risks to deployed personnel, particularly in view of the multiple exposures, chemicals, toxic end points, and/or sources of uncertainty likely to be involved. The italicized portions focus on the critical need for *comparative* risk analysis that would allow commanders to make trade-off decisions concerning uncertain risks in the context of potentially competing goals, ranging from combat success to preventive public or occupational health, that might differ in urgency. Such trade-offs should reflect *unbiased* assessments of net risk associated with alternative courses of action (NRC 2000). The same recommendation appears in an earlier NRC report, *Science and Judgment in Risk Assessment* (NRC 1994), which states that "decisions involving risk-trading or priority setting ... should take into account information on uncertainty in quantities being ranked so as to ensure that such trades do not increase expected risk."

THE ARMY'S RISK-ASSESSMENT GUIDANCE FOR DEPLOYMENT

Description

Technical Guide 248 (TG-248)

TG-248 (USACHPPM 2001) outlines the processes and tools that could be used to evaluate and communicate all categories of occupational and environmental health (OEH) and endemic disease (ED) hazards in accordance with the military operational risk-management (ORM) process discussed in Chapter 1. TG-248 focuses on the first two steps of the ORM process, identifying OEH/ED hazards and assessing the threat they pose to the mission in terms of their probability and severity. TG-248 was designed to enable preventive-medicine personnel to express the risks from each OEH/ED hazard in the same metric used for other more traditional military hazards (e.g., enemy forces, mechanical problems) as well as other OEH/ED hazards so that decision makers can make rational comparisons of the various risks faced during deployment and make decisions about courses of action.

The approach used by the U.S. Army Center for Health Promotion and Preventive Medicine (USACHPPM) is necessarily different from those used for traditional occupational or environmental risk assessments because military decision making must consider mission impact in addition to individual health risk. Thus, TG-248 facilitates classification of OEH/ED hazards using a standard component of military ORM—the risk-assessment matrix (see Table 2-1). The risk-assessment matrix is a qualitative classification scheme that reflects four categories of "severity" of risk to a military mission and five categories of "probability" with regard to one or more military assets and/or soldiers. These two dimensions are combined to comprise 20 cells that are separated into four qualitative categories of mission-related risk: extremely high, high, moderate, and low. These risk categories pertain specifically to the qualitative likelihood of mission success, which refers to mission-specific military goals including, but not limited to, the minimization of health risks to deployed military personnel.

Technical Guide 230 (TG-230)

TG-230 (USACHPPM 2001a) adapts the generalized framework of TG-248 and proposes a specific process to evaluate the chemical subset of

TABLE 2-1 Risk-Assessment Matrix

		Probability				
		Frequent	Likely	Occasional	Seldom	Unlikely
Severity		A	B	C	D	E
Catastrophic	I	E	E	H	H	M
Critical	II	E	H	H	M	L
Marginal	III	H	M	M	L	L
Negligible	IV	M	L	L	L	L

Definitions

Hazard Severity

Catastrophic (I): Loss of ability to accomplish the mission or mission failure. Death or permanent disability. Loss of major or mission-critical system or equipment. Major property (facility) damage. Severe environmental damage. Mission-critical security failure. Unacceptable collateral damage.
Critical (II): Significantly degraded mission capability, unit readiness, or personal disability. Extensive damage to equipment or systems. Significant damage to property or the environment. Security failure. Significant collateral damage.
Marginal (III): Degraded mission capability or unit readiness. Minor damage to equipment or systems, property, or the environment. Injury or illness of personnel.
Negligible (IV): Little or no adverse impact on mission capability. First aid or minor medical treatment. Slight equipment or system damage, but fully functional and serviceable. Little or no property or environmental damage.

Risk Levels

E – Extremely high risk: Loss of ability to accomplish the mission if threats occur during mission. A frequent or likely probability of catastrophic loss (IA or IB) or frequent probability of critical loss (IIA) exists.
H – High risk: Significant degradation of mission capabilities in terms of the required mission standard, inability to accomplish all parts of the mission, or inability to complete the mission to standard if threats occur during the mission. Occasional to seldom probability of catastrophic loss (IC or ID) exists. A likely to occasional probability exists of a critical loss (IIB or IIC) occurring. Frequent probability of marginal losses (IIIA) exists.
M – Moderate risk: Expected degraded mission capabilities in terms of the required mission standard will have a reduced mission capability if threats occur during mission. An unlikely probability of catastrophic loss (IE) exists. The probability of a critical loss is seldom (IID). Marginal losses occur with a likely or occasional probability (IIIB or IIIC). A frequent probability of negligible (IVA) losses exists.
L – Low risk: Expected losses have little or no impact on accomplishing the mission. The probability of critical loss is unlikely (IIE), while that of marginal loss is seldom (IIID) or unlikely (IIIE). The probability of a negligible loss is likely or less (IVB through IVE).

Hazard Probability

Frequent (A): Occurs very often, continuously experienced.
Likely (B): Occurs several times.
Occasional (C): Occurs sporadically.
Seldom (D): Remotely possible; could occur at some time.
Unlikely (E): Can assume will not occur, but not impossible.

Unit Status

Black: Unit requires reconstitution. Unit below 50% strength.
Red: Combat ineffective. Unit at 50-69% strength.
Amber: Mission capable, with minor deficiencies. Unit at 70-84% strength.
Green: Mission capable. Unit at 85% strength or better.

Source: TG-230 and U.S. Army Field Manual 3-100.12.

OEH/ED hazards. In the first step of this process, chemical hazards are identified through available intelligence data, field sampling, and/or exposure modeling. Potential chemical hazards are then prioritized on the basis of whether they pose no threat, a health threat, or a medical threat. Health threats are hazards that could result in adverse health effects in an individual. Medical threats are those health threats that have the potential to render a field unit ineffective for combat or for other mission-related activities. Threats of chronic or delayed disease (e.g., cancer, liver disease, or kidney disease) are categorized as *threats of concern to the command,* which are generally considered health threats, but on occasion could be considered medical threats. Each chemical is categorized by comparing measured or predicted concentrations of that chemical with its most relevant military exposure guideline (MEG). MEGs are estimated chemical concentrations above which certain types of adverse health effects might begin to occur in individuals within the exposed population after a continuous, single exposure of specified duration. They were designed to address the wide variety of exposure scenarios that could be encountered during deployment, ranging from catastrophic release of a large amount of chemical to regional pollution. MEGs were developed by modifying the existing exposure standards set by other agencies for application to the military context. The process by which MEGs were derived is described in Reference Document 230 (RD-230) (USACHPPM 2001b). That process is evaluated by the subcommittee in Chapter 5.

Chemical threats are further categorized by using a hazard severity ranking chart that refers to the four categories of severity defined by the risk-assessment matrix. The severity ranking charts provided in TG-248 and TG-230 also incorporate specific ranges of probabilities of symptoms grouped by severity category (see Tables 2-2 and 2-3). Chemical risks are finally classified using the risk levels defined in the military risk-assessment matrix. The risk levels correspond to "unit status" levels that are color coded. Unit status refers to effective unit strength expressed as a percentage (see Table 2-4). In both guidance documents, these probabilistic unit strength levels are related to hazard severity levels and hazards probability categories, but those relationships are made quantitatively explicit in TG-230 insofar as that document defines the hazard probability categories discussed above in terms of corresponding troop exposure probability ranges (see Table 2-5). A confidence level is then assigned to the risk estimate by using criteria outlined in TG-248 (see Table 2-6). That judgment is made by considering key sources of uncertainty associated with the risk assessment, such as the quality of the field sampling data and understanding of the exposure conditions.

TABLE 2-2 TG-230 Chemical Hazard Severity Ranking Chart for Military Deployments

Magnitude of Chemical Concentration			Associated Health Outcome Attributable to Exposure			
Air	Water	Soil	(General)[a]	Onset of Symptoms	Hazard Severity Rank	Hazard Type
<1-year MEG or <14-day MEG	<MEG	<MEG	No cases of illness or noncancer disease and <1 cancer case in 10,000	AFTER THE MISSION	NONE	NO HEALTH THREAT
≥1-year MEG or ≥14-day MEG but ≤1-24 hour min-MEG	≥MEG that is not based on TB MED 577[b]	≥MEG[c]	0-10% of personnel might develop illness or chronic disease		NEGLIGIBLE	HEALTH THREAT
≥1-year MEG or ≥14-day MEG but >1-24 hour min-MEG			0-10% of personnel might develop mild illness or temporary irritation	DURING THE MISSION		
>1-hour min-MEG but ≤1-hour sig-MEG	≥MEG that is based on TB MED 577[b]	—[c]	>10% of personnel might experience mild illness and irritation;		MARGINAL	MEDICAL THREAT
			0-10% of personnel might develop more severe illness that begins to impair functional capabilities			

(Continued)

TABLE 2-2 Continued

Magnitude of Chemical Concentration			Associated Health Outcome Attributable to Exposure			
Air	Water	Soil	(General)[a]	Onset of Symptoms	Hazard Severity Rank	Hazard Type
>1-hour sig-MEG but ≤1-hour sev-MEG	—[b]	—[c]	10-25% of personnel might experience severe illness or irritation and more noticeable degradation of performance capabilities; Other personnel will, at least, suffer some mild effects	During the mission	CRITICAL	MEDICAL THREAT
>1-hour sev-MEG	—[b]	—[c]	>25% of personnel might experience severe, incapacitating effects; Fatalities will begin to occur just above the sev-MEG for air exposur and will increase as concentrations increase		CATASTROPHIC	

[a]Percentages are very uncertain and will vary by chemical and by other confounding factors.
[b]Concentrations greater than the MEG *might* result in hazard severity from marginal to catastrophic if certain chemicals are present in large enough quantities and there is sufficient consumption. Additional information in the "Notes" column of the MEG tables should be evaluated regarding effects of higher levels of exposure.
[c]Soil is unlikely to represent a hazard that would yield a medical threat. Additional information in the "Notes" column of the MEG tables should be evaluated for data regarding higher levels of exposure.
Abbreviations: min-MEG, minimal effects level; sig-MEG, significant effects level; sev-MEG, severe effects level.
Source: Modified from USACHPPM 2002a.

TABLE 2-3 TG-248 Hazard Severity Ranking Chart for Military Deployments

	Nature of Individual Health Effects Associated with Exposures Near the Guideline			
	Symptoms Occurring After the Mission	Symptoms Occurring During the Mission		
Percent of Exposed People to Exhibit Symptoms (Attack Rate)	Chronic or Permanent Injury or Disease[a]	Mild Illness or Temporary Irritation[b]	Injury or Illness That Impairs Functional Capabilities	Incapacitation or Death
>50%	Marginal	Critical	Catastrophic	Catastrophic
31-50%	Negligible	Marginal	Critical	Catastrophic
10-30%	Negligible	Marginal	Marginal	Critical
<10%	Negligible	Negligible	Marginal	Critical

[a]For example, cancer.
[b]Reversible, short-term, nuisance.
Source: Modified from USACHPPM 2001.

TABLE 2-4 TG-230 Risk Level Definitions

Risk Level	Defined Consequence[a]	Unit Status[b,c]
Extremely high	Expected loss of ability to accomplish the mission	Black (unit requires re-constitution) Unit below 50% strength
High	Expected significant degradation of mission capabilities in terms of the required mission standard, inability to accomplish all parts of the mission, or inability to complete the mission to standard if hazards occur during the mission	Red (combat ineffective) Unit at 50-69% strength
Moderate	Expected degraded mission capabilities in terms of the required mission standard will have a reduced mission capability if hazards occur during mission	Amber (mission capable, with minor deficiencies) Unit at 70-84% strength
Low	Expected losses have little or no impact on accomplishing the mission	Green (mission capable) Unit at 85% strength or better

[a]Field Manual 100-14 (U.S. Department of the Army 1998)
[b]Field Manual 3-100.12 (U.S. Department of the Army 1997)
[c]The unit rates provided under "Unit Status" are to be determined by the commander. Charts similar to the example hazard probability and severity ranking charts presented in Tables 2-1, 2-2, and 2-3 should be aligned with the acceptable risk levels provided by the commander.
Source: USACHPPM 2000a.

Finally, using all the information at hand, the threat category is re-evaluated in terms of whether the chemical poses no threat, a health threat, a threat of concern to the command, or a medical threat. The purpose of this final step is to provide perspective on which hazards pose greater operational threats when comparing threats that have similar risk estimates.

TABLE 2-5 TG-230 Chemical Hazard Probability Ranking Chart for Military Deployments

Percent of Personnel That Will Experience Exposures to Concentrations Equal To or Greater Than the MEG[a]				
<10%	10-25%	25-50%	50-75%	>75 %
Unlikely	Seldom	Occasional	Likely	Frequent

[a]Determination of the percent of personnel exposed to a chemical or mixture specifically above a guideline level can be based on modeling, gridding, or generalized assumptions.
Source: USACHPPM 2002a.

TABLE 2-6 TG-230 Example Criteria for Assigning Confidence Levels

Confidence Level	Criteria
High	Sampling data quality is good Field activity patterns are well known True exposures are reasonably approximated Knowledge of the symptoms of hazard exposure relative to guideline is well known No important missing information The predicted health outcome is plausible or already demon strated
Medium	Field data quality is good Field exposures are likely to be overestimates of true exposures due to incomplete data coverage relative to actual expo sure durations Detailed information is lacking regarding true personnel activ ity patterns in the field Symptoms are well known for each individual hazard, but some scientific evidence suggests that the combined effects of all hazards may exacerbate symptoms Predicted health outcome is plausible
Low	Important data gaps and/or inconsistencies exist Exposure conditions are not well defined Field personnel activity patterns are basically unknown Predicted health outcome is not plausible because it is not consistent with real-world events/experience

Source: USACHPPM 2002a.

Evaluation[1]

The risk-management framework presented in TG-230 and TG-248 reflects much of the guidance and recommendations provided by the NRC (2000) and IOM (2000). DOD, and in particular USACHPPM, is to be commended for developing a risk-management framework that implements the recommendations contained in those reports. The generalized framework in TG-248 also comprises innovative features that are suited for practical use in the field. The framework attempts a quantitative implementation of a matrix approach to assess risk levels that are implied by different categories of hazard severity and hazard probability. It uses familiar categories previously defined in DOD's overall approach to military risk management, as described in field manuals FM 100-14 and FM 3-100.12. This

[1]Minor errata and inconsistencies found in TG-248, TG-230, and RD-230 are discussed in Appendix A.

framework is particularly well-suited to convenient characterization and evaluation of individual health or medical threats and simple comparisons of those individual threats, provided comparable levels of uncertainty are involved.

When applied to assess chemical risks in TG-230 (Table 2-2), the framework enables the use of a simple, tabular approach to characterize exposure scenarios chemical by chemical in terms of a corresponding categorical risk level that in turn refers to a corresponding range of unit strength (expressed as a percentage). The approach also incorporates traditional procedures used to identify noncompliance with occupational safety and health guidelines pertaining to chemical exposures that are expected to produce health effects (i.e., hazards of "negligible" severity). The latter involves the use of MEGs to assess the significance of field exposures to specific chemicals.

Overall, the subcommittee found that TG-248 provides a reasonable categorization process for assessing OEH/ED hazards within the context of ORM. The process allows for predicting impacts on missions and making reasonable comparisons between the potential hazards, be they hazards to equipment, troops, or some other facet of deployment. However, the chemical-risk guidance (i.e., the MEGs) in TG-230 is inconsistent with the intent of TG-248 and could lead to mischaracterization of the significance of chemical risks in comparison with other deployment risks. This potential for mischaracterization is the result of USACHPPM's attempt to use one set of exposure guidelines (i.e., MEGs) for the dual purposes of "characteriz[ing] the level of *health* and *mission* risks associated with identified or anticipated exposures to chemicals in the deployment environment" (USACHPPM 2001, p.1; italics added). These conflicting purposes could lead to different interpretations and uses of the MEGs. For example, RD-230 states the following:

> In some cases, exposures greater than the MEG can induce immediate adverse health effects, in other cases exposures greater than the MEG simply indicate that there is an increased likelihood that a health problem could arise either during or post deployment ... In general, environmental concentrations equal to, or slightly greater than the specified MEG are expected to result in the specified type and degree of health effects in none to a small portion of individuals in the exposed military population. In some cases, MEG represents a purely "protective" level where health effects should not be observed at all.

The subcommittee summarized the differing goals of mission and health risk assessment in Table 2-7. The table shows that the parameters for achieving each of those goals are quite different from each other on all levels. Those differences make it extremely difficult for one set of guidance values (MEGs) to adequately address both sets of goals.

The risk-assessment matrix used in TG-230 is a device for categorizing chemical risks in terms of mission impact (e.g., mission capable, combat ineffective) so that chemical hazards can be weighed against and compared with other hazards to missions (e.g., mechanical failures, weather). To make comparative assessments among all potential deployment hazards, it is important to assess all hazards in terms of their potential impact on unit strength. Assessment of those hazards requires an understanding of the chemical exposure levels at which casualties that would render the unit ineffective might begin to occur. MEGs are inappropriate for making this type of assessment because they are *health protective* values that provide an estimated threshold at which health effects might begin to occur. Those threshold concentrations could be several orders of magnitude below those that would be anticipated to produce enough casualties to compromise a mission. Thus, mission risks characterized using the MEGs would not be comparable to the risk levels assigned to other operational hazards using the risk-assessment matrix. See example in Box 2-1. The subcommittee believes that separate sets of chemical risk assessment guidance are needed for assessing mission and health risks. Those assessments should be presented simultaneously to help decision makers balance the necessary trade-offs between the mission's needs and potential health impacts presented by the deployment mission under consideration.

Mission Risk Assessment

To assess mission risks, it will be necessary for USACHPPM to develop a set of unbiased, predictive estimates of casualties that might occur if the unit is exposed to a particular concentration of a chemical. The subcommittee has termed such values chemical casualty estimating guidelines (CCEGs) and recommends that they be used instead of MEGs to characterize mission risks using the ORM categorization scheme. CCEGs would be media and duration-specific chemical concentrations expected to cause health impairments that degrade the performance of enough individuals to reduce unit strength, also known as medical threats. As discussed in more detail in Chapter 4, CCEGs cannot be established from existing exposure

TABLE 2-7 Characteristics Associated with the Major Goals of TG-248 and TG-230

	Health Risk Assessment	Mission Risk Assessment
Goal	To assess impacts on individual soldier health; requires the use of *protective* exposure values	To predict impacts of health risks on the mission; requires the use of *predictive* casualty estimates
Effects	Short- and long-term effects	Primarily short-term effects
Length of exposure	Long-term exposure	Short-term exposure
Situation	More like occupational/environmental (OSHA, EPA)	More like short-term emergency planning
Availability of data	More likely to have data available to assess exposure	More qualitative assessment of exposure; relies more on subjective judgment
Availability of time	More time to assess	Decisions must be made quickly
Exposure assessment	Assess proportion likely to receive exposure in excess of MEGs	Assess proportion likely to receive any mission-compromising level of exposure
Number of chemicals	Many of concern	Limited number of concern
Likelihood that effect(s) will occur	Lower	Higher
Confidence in estimated exposure(s)	Higher	Lower

standards, but should be derived by conducting independent evaluations of each chemical and developing exposure-response and population-response information to make casualty estimations. Although this will be a significant undertaking, the number of chemicals and the scope of the data needed for evaluation are expected to be far less than those that were required to set the MEGs.

USACHPPM should further prioritize the general risks posed by specific media of exposure, including air, soil, and water, and the available risk-management options for various durations of missions. For example, water contamination could be a serious risk for long-term missions, espe-

BOX 2-1 Scenario: Securing a River Crossing

The mission is to secure a major river crossing. If the enemy succeeds in crossing the river, thousands of lives will be lost. The commander has the choice of two routes to get to the crossing. One is a very rugged road that poses a threat to the mission because of potential vehicle roll-over, mechanical failure, and other hazards related to terrain impacts on vehicles. The other route is paved but goes by a chemical plant that has structural damage and is believed to be leaking chemical X. Chemical X can cause irritation of the mucous membranes and respiratory system, headache, and nausea that could impair the functional capabilities of the troops. In addition, chemical X is associated with potential long-term effects.

To decide which route to take, the commander needs to know the potential terrain risks and chemical risks to the unit. For the terrain risks, the commander will be provided with an assessment of whether enough vehicles will be disabled during the rugged terrain crossing such that an insufficient number of troops and equipment will be able to reach and secure the river crossing. For a comparable assessment of chemical risks, a prediction is needed of whether the exposure incurred by passing the chemical plant will disable enough troops to the extent that they could not perform their duties. However, using the current set of guidance, it would be impossible to get a chemical assessment comparable to the terrain assessment, because the relevant MEG for chemical X in TG-230 will not be a casualty estimate but a health protective guideline. The short-term MEGs for chemical X will define a level at which respiratory irritation, headache, and nausea would begin to occur, and using them as benchmarks in conjunction with the risk-assessment matrix would result in overestimating the risk that chemical X poses to the mission. In addition to the assessment of risks to the mission, the commander will also need to be informed of the long-term health risks posed by chemical X.

cially when establishing a base of operations. However, water contamination generally is the easiest to mitigate by avoidance, treatment, or use of alternative sources. Air contamination is probably the most important consideration for short-term missions. Short-duration releases of chemical warfare agents or deliberate releases of toxic industrial chemicals could have immediate acute effects. A general example of risks prioritization on the basis of duration of exposure and mitigation options is provided in Table 2-8. USACHPPM should use a similar scheme to establish a set of criteria for CCEGs. A preliminary characterization of those guidelines is provided in Box 2-2.

Important issues to consider in developing CCEGs include the following:

- Risk should be defined as an unbiased or "best" estimate. When an integrated quantitative approach is used to estimate risk and its uncertainty, an appropriate corresponding attribute of a risk or casualty number to esti-

TABLE 2-8 General Prioritizations of Exposure Routes in Relation to Exposure Duration and Possible Mitigation

Exposure Duration	Routes of Exposure[a] and Mitigation		
Short mission	Air	Soil	Water
	Respirator	PE (booties)	Avoid
Long mission	Water	Air	Soil
	Treatment	Filtered shelter	Avoid

[a]In order of decreasing risk, left to right.

mate would be its expected value, or "population mean," defined mathematically as the arithmetic average of all possible likelihood-weighted values.[2]

- TG-230 makes it clear that "unit strength" should refer not only to directly affected personnel but also to individuals affected to a lesser extent

[2]"Likelihood" weights in this context mean probabilities from a distribution that reflects uncertainty in a true but unknown value that must be estimated. Expected values have two key properties that do not generally apply to other measures of central tendency, such as medians or modes. First, the arithmetic mean of a random sample of observed values of any uncertain variable always provides an unbiased (i.e., not systematically error-prone) estimate of the (unknown) expected value of that variable. Second, the expected value of any function of uncertain but statistically independent input variables can always be estimated conveniently (at least to a first order of approximation) by calculations that involve only estimates of the expected value of each separate input variable. If a function of independent input variables is "linear" (i.e., involves only the sum, difference, product, and/or ratio of those variables), then the expected value of the function always exactly equals the function of the expected values of the input variables. For example, using $e(z)$ to denote "the expected value of an estimated variable z," if a total number (n) of casualties is estimated as the product of independent and uncertain variables representing concentration (c), potency (q), and number (x) of potentially exposed personnel, then it will always be true that $e(n) = e(c) \times e(q) \times e(x)$, regardless of the shapes of the distributions that characterize uncertainty in c, q, and x, respectively.

These two useful properties of expected values can be applied conveniently to compare alternative courses of action (say, COA1 vs COA2) expected to generate two corresponding numbers (N1 vs N2) of casualties. A reasonable and consistent rule often used for this type of decision is to always reject COA2 if $e(\text{N2} - \text{N1}) > 0$. This rule by definition minimizes the expected number of predicted casualties (Raiffa 1970). By virtue of the expected-value properties mentioned above, this rule always can be re-expressed validly in a form that conveniently involves only the expected value of each separate casualty prediction—namely, always reject COA2 if $e(\text{N2}) > e(\text{N1})$. Although more elaborate decision rules can be used to account for the shape of the distribution of uncertainty in the difference (N2 - N1) relative to a specified set of risk-aversion preferences, complex rules of this type typically require quantitative uncertainty analysis methods likely to be impractical for most current military operational risk-management contexts.

BOX 2-2 CCEG Characteristics

- Include chemicals likely to be encountered in sufficient quantities to degrade mission effectiveness.
- Include health effects that are manifested within minutes, hours, or several days that could immediately affect the functioning of troops (e.g., loss of cognitive ability, loss of visual acuity, significantly reduced cardiopulmonary functioning, muscular weakness) performing a mission. Does not include long-term health effects (e.g., cancer).
- Consider exposure time frames of hours, days, and weeks, rather than months or years.
- Relate to the military population, which includes generally healthy adult men and women with typical variations in genetic susceptibilities.
- Provide exposure-response and population-response information, insofar as possible. Include concentrations likely to cause effects in humans, along with a description of the severity and incidence expected. This information would enable chemical threats to be weighed in comparison to other mission threats (e.g., Table 3-1 in TG-230 would be more useful).
- Provide guidance primarily for the air exposure pathway, because troops have no choice but to breathe the air (except when gas masks are used). Theoretically, the water pathway might influence CCEGs, but the availability of alternative sources of water makes it relatively less important. Water exposure scenarios of special concern should be identified and addressed. Soil also deserves some consideration, but is unlikely to be a significant source of exposure.

and support personnel required to tend to directly affected personnel. The multiplication of hazard severity probabilities (i.e., illness likelihood ranges) by hazard probability ranges clearly is not intended to yield corresponding defined-target ranges of "fractional unit incapacitation" (i.e., the opposite of unit strength, as defined in TG-230). Therefore, USACHPPM must consider developing CCEGs that provide an assessment of overall unit incapacitation with increasing exposure levels. (TG-230 also mentions that severe toxic effects will lead to increased medical support requirements for affected personnel, but it was not clear to the subcommittee how that support necessarily would reduce unit effectiveness, insofar as required medical support personnel would be performing their intended function.)

- The need for a categorical confidence-level scheme (high vs medium vs low) in TG-248 and TG-230 should be reconsidered. The assignment of confidence levels is not a recommendation in the ORM process presented in Field Manual 3-100.12 (DOD 2001), so other operational risks will not be assigned confidence levels. It is unclear how decision makers are to interpret a specified confidence level, particularly a low one, when trying to balance competing operational risks. A low confidence in the risk characterization gives no indication of whether actual risk might be higher or lower than the predicted risk. Furthermore, it is unclear how confidence level description for chemical risks would be of value to decision makers if confidence levels are not assigned to other operational risks.

Chapter 4 presents more specific guidance for developing CCEGs and discusses how the guidelines should be applied.

Health Risk Assessment

Another goal of TG-230 is to provide force health protection, with the understanding that mission success has primacy over some health risks that might be considered unacceptable under less hazardous conditions. With some modifications, MEGs can be used to fulfill that goal. To that end, MEGs would be concentrations of chemicals in air, water, and soil that can be used to estimate the potential impact of field exposures on soldier health during deployments. A preliminary characterization of MEGs is presented in Box 2-3. MEGs would be used to determine the appropriate management actions that could be taken to avoid or mitigate risks. Depending on the particulars of the deployment scenario, commanders could decide whether the benefits of the mission outweigh the possible health risks to individual soldiers. In cases where commanders decide to accept the health risks to soldiers, MEGs could be used to determine what kinds of health-management actions to take, such as documenting exposures in soldiers' records, conducting additional environmental sampling, or conducting follow-up health monitoring. Chapter 5 provides a more detailed description of how MEGs should be derived and applied.

BOX 2-3 MEG Characteristics

- Include a large number of chemicals likely to be present in deployments.
- Include concerns over longer-term health of individuals, but recognizing that exposures at these levels would have no to minimal impact on immediate missions.
- Include virtually all exposure durations from 1 hour to 1 year.
- Relate to the military population.
- Indicate protective levels (i.e., levels assumed to represent no adverse effects or very low risk) for the exposure durations of interest.
- Provide guidance for management actions when MEGs are exceeded.

RECOMMENDATIONS

- TG-230 should be revised to provide separate guidance on assessing chemical hazards for the purposes of mission and health risk assessment.
- For the purposes of mission risk assessment, risks should be evaluated within the context of mission success. The subcommittee recommends the development of chemical casualty estimating guidelines (CCEGs) to provide an appropriate basis for comparison with other mission hazards. CCEGs would be media and duration-specific chemical concentrations expected to cause health impairments that debilitate the performance of enough individuals to significantly reduce unit strength and effectiveness. They would be predictive values that provide unbiased quantitative exposure-response and population-response information that enables commanders to compare the risks from chemical threats to those from other mission threats (e.g., combat casualties, logistical problems) using the same metric. The goal is to provide reasonably accurate estimates of impacts on unit strength. Chapter 4 provides guidance on how to derive CCEGs.
- CCEGs should be developed for a subset of the chemicals for which MEGs have already been derived. Chemicals should be selected on the basis of their potential as immediate medical threats to missions. Warfare agents and high-production-volume industrial chemicals with high toxicological potency are the most likely candidates. Inhaled volatile chemicals or toxic particulate matter also are likely to fall into that category, although chemical exposures by ingestion and skin contact should likewise be considered.
- The need for assigning confidence levels to mission-risk estimates should be reconsidered.
- For the purposes of health protection, chemical risks should be assessed independently of operational goals. USACHPPM's current set of military exposure guidelines (MEGs) are health-protective values that, with some modification, could fulfill that need. MEGs should be used to define risk-management actions that can be taken to avoid or mitigate potential health risks. Chapter 5 expands on this recommendation.
- Mission-risk and health-risk information should be provided to decision makers simultaneously, so that risks can be balanced explicitly and appropriate management actions can be taken.
- As guidance on applying TG-248 to other OEH/ED hazards (e.g., radiological and biological hazards) is developed, USACHPPM should consider combining all of the guidance into a single document to facilitate consideration of cumulative risks from all environmental hazards.

REFERENCES

DOD (U.S. Department of Defense). 2001. Risk Management. Multiservice Tactics, Techniques and Procedures. FM 3-100.12. MCRP 5-12.1C. NTTP 5-03.5. AFTTP(I) 3-2.34. Air Land and Sea Application Center. U.S. Army Training and Doctrine Command, Fort Monroe, VA; Marine Corps Combat Development Command, Quantico, VA; Navy Warfare Development Command, Newport, RI; and Headquarters Air Force Doctrine Center, Maxwell Air Force Base, AL. [Online]. Available: http://www.adtdl.army.mil/cgi-bin/atdl.dll/query/download/FM+3-100.12 [accessed Dec. 2, 2003].

IOM (Institute of Medicine). 2000. Protecting Those Who Serve: Strategies to Protect the Health of Deployed U.S. Forces. Washington, DC: National Academy Press.

NRC (National Research Council). 1983. Risk Assessment in the Federal Government: Managing the Process. Washington, DC: National Academy Press.

NRC (National Research Council). 1994. Pp. 161-244, 516-518 in Science and Judgment in Risk Assessment. Washington, DC: National Academy Press.

NRC (National Research Council). 2000. Strategies to Protect the Health of Deployed U.S. Forces: Analytical Framework for Assessing Risks. Washington, DC: National Academy Press.

Raffia, H. 1970. Decision analysis: introductory lectures on choices under uncertainty. 2nd Ed. Harvard University Press, Cambridge, MA.

USACHPPM (U.S. Army Center for Health Promotion and Preventive Medicine). 2001. Guide for Deployed Preventive Medicine Personnel on Health Risk Management. Technical Guide 248. U.S. Army Center for Health Promotion and Preventive Medicine. August 2001. [Online]. Available: http://chppm-www.apgea.army.mil/ deployment/ [accessed November 25, 2003]

USACHPPM (U.S. Army Center for Health Promotion and Preventive Medicine). 2002a. Chemical Exposure Guidelines for Deployed Military Personnel. Technical Guide 230. U.S. Army Center for Health Promotion and Preventive Medicine. January 2002. [Online]. Available: http://chppm-www.apgea.army.mil/deployment/ [accessed November 25, 2003]

USACHPPM (U.S. Army Center for Health Promotion and Preventive Medicine). 2002b. Chemical Exposure Guidelines for Deployed Military Personnel. A Companion Document to USACHPPM Technical Guide (TG) 230 Chemical Exposure Guidelines for Deployed Military Personnel. Reference Document (RD) 230. U.S. Army Center for Health Promotion and Preventive Medicine January 2002. [Online]. Available: http://chppm-www.apgea.army.mil/deployment/ [accessed November 25, 2003]

U.S. Department of the Army. 1997. Operational Terms and Graphics. Field Manuel No. FM 101-5-1/MCRP 5-2A. Headquarters, Department of Army, U.S. Marine Corps, Washington, DC. September 30, 1997.

U.S. Department of the Army. 1998. Risk Management, Field Manual No. 100-14. U.S. Department of the Army, Washington, DC. April 23, 1998.

3

Review of Key Concepts, Assumptions, and Decisions Made in Developing TG-248, TG-230, and RD-230

This chapter provides an overview of some of the general concepts discussed and key assumptions and decisions made by the U.S. Army Center for Health Promotion and Preventive Medicine (USACHPPM) in developing Technical Guide 248 (TG-248), Technical Guide 230 (TG-230), and Reference Document 230 (RD-230), emphasizing the issues identified in the subcommittee's statement of task. The subcommittee evaluated the following aspects of the Army's guidance: the use and adaptation of pre-existing exposure guidelines for deployment purposes; population susceptibilities; exposure factors; acceptable lifetime cancer risk; immediate and long-term health effects; aggregate exposure and cumulative risk; exposure assessment; and the utility of the guidance for decision makers.

USE OF PRE-EXISTING EXPOSURE GUIDELINES

Military exposure guidelines (MEGs) were developed by USACHPPM for contaminants in air, water, and soil. They were derived by reviewing the guidelines of other agencies (e.g., the U.S. Environmental Protection Agency [EPA], the Occupational Safety and Health Administration [OSHA]), selecting the most relevant guidelines on the basis of a hierarchical scheme, and modifying the chosen guidelines for military use. The drawback of this approach is that the existing guidelines were designed to protect different populations (e.g., the general population, workers) and were intended for different settings (e.g., ambient exposures, workplace, accidental releases), which made it necessary for USACHPPM to adjust the values to make them relevant to the military population in the deployment setting. Problems with using pre-existing guidelines and adjusting them for deployment purposes are described below.

Procedures for Developing Noncancer and Cancer Health Assessments

The following procedures typically are used by regulatory and other agencies to establish health-protective exposure guidelines and therefore form bases of the MEGs.

Noncancer Assessments

Most noncancer assessments begin by selecting a no-observed-adverse-effect level (NOAEL) from experimental data and adjusting and extrapolating that value by applying factors to account for uncertainties related to exposure duration, varying levels of susceptibility among humans or between species when animal data are being used, and other facets of the data. Typically, the adverse effect having the lowest NOAEL in the most sensitive species for which data are available is chosen as the critical toxicity end point for derivation of the guideline. The assumption is that if the population is protected from that adverse effect, it will also be protected from the other adverse effects observed at higher concentrations. NOAELs can be determined by identifying the lowest NOAEL from a single critical study or by doing a benchmark dose analysis and selecting the mathematical result to use as a surrogate NOAEL. Data from the selected study or studies of interest are typically transformed to a product of concentration and time (i.e., $C \times t$) to account for differences between the exposure duration used in the study or studies and the duration for which the health-protective guideline is being established.

The NOAEL is adjusted by the use of uncertainty factors (UFs). These factors are applied to account for uncertainties in extrapolating experimental animal data to humans (interspecies differences) or variable susceptibilities in the human population (intraspecies differences); to represent the expected ratio of the lowest-observed-adverse-effect level (LOAEL) to NOAEL when a LOAEL is used instead of a NOAEL; to account for uncertainty in predicting chronic exposure effects on the basis of subchronic exposure studies; and to provide a margin of safety when the database is incomplete (sometimes referred to as a modifying factor). Standardized UFs are derived from literature comparisons (e.g., comparing results from a subchronic and chronic study to estimate what value might be applied to a subchronic result to conservatively predict the chronic result). UFs are not statistically derived indicators of uncertainty, and most are used to account for missing information (e.g., only subchronic data exist, but a chronic exposure value is needed). Thus, each application of a UF indicates that key information required to predict a chemical-specific toxic end point is not available; it must instead be addressed by using a default estimate of the potential magnitude of the corresponding impact of that factor on the likelihood of the end point of concern. The factor sometimes applied to account for intraspecies differences is more appropriately referred to as a variability factor, insofar as it is applied to address interindividual heterogeneity and not uncertainty.

Most UFs are either 10 (a log), 3 (half a log), or 1 (no UF). EPA uses all of the UFs described above, as needed, up to a maximum of 10,000 for reference doses (RfDs) and 3,000 for reference concentrations (RfCs). The maximum for the RfC is lower because interspecies differences are handled by using a combination of a dosimetric extrapolation and a maximum UF of 3 for pharmacodynamic differences. The minimal risk levels (MRLs) of the Agency for Toxic Substances and Disease Registry (ATSDR) include all of the UFs except the duration adjustment, because guidelines are developed for several durations.

The use of UFs differs among the existing exposure guidelines, leading the current MEGs to vary in their conservatism. Table 3-1 presents the UFs that underlie some of the MEGs. The table shows that some of the applied UFs are not relevant to the deployed population, as is the case for the UFs for phosphine. In other cases, it could not be determined whether UFs were used, which makes it impossible to assess the level of protection provided. USACHPPM attempted to make adjustments to account for the differences in characteristics between the deployed and general population, but the subcommittee found the adjustments were not sufficient to ensure that the resulting MEGs provide comparable levels of protection among chemicals. Chapter 5 provides a more detailed description of the procedures used by various organizations to derive their exposure guidelines, a summary of adjustments applied by the military, and recommendations for making improvements in the MEGs.

TABLE 3-1 Implied Underlying Basis of 1-Hour Air MEGs for Selected Chemicals

Chemical	1-Hour Air-MEG	Basis	Species	U_H	V	U_L	U_D
Acrolein	Significant	AEGL-2 (P)	Human	1	3	1	1
	Severe	AEGL-3 (P)	Animal	3	3	1	1
Acrylonitrile	Significant	ERPG-2	Human	NS	NS	NS	NS
	Severe	ERPG-3	Animal	NS	NS	NS	NS
Arsine	Significant	AEGL-2 (F)	Animal	10	3	1	1
	Severe	AEGL-3 (F)	Animal	10	3	1	1
Bromine	Significant	AEGL-2 (P)	Human	1	3	1	1
	Severe	AEGL-3 (P)	Animal	3	3	1	1
Chlorine	Significant	AEGL-2 (I)	Human	1	1	1	1
	Severe	AEGL-3 (I)	Animal	3	3	1	1
Diborane	Significant	AEGL-2 (F)	Animal	3	3	1	1
	Severe	AEGL-3 (F)	Animal	3	3	1	1
Formaldehyde	Significant	ERPG-2	Human	NS	NS	NS	NS
	Severe	ERPG-3	Animal/Human	NS	NS	NS	NS
Hydrogen chloride	Significant	AEGL-2 (I)	Animal	3	3	1	3[a]
	Severe	AEGL-3 (I)	Animal	3	3	1	1
Hydrogen cyanide	Significant	AEGL-2 (F)	Animal	2	3	1	1
	Severe	AEGL-3 (F)	Animal	2	3	1	1
Hydrogen fluoride	Significant	AEGL-2 (I)	Animal	3	3	1	1

	Severe	AEGL-3 (I)	Animal	1	3	1	2[b]
Hydrogen sulfide	Significant	AEGL-2 (I)	Animal	3	3	1	1
	Severe	AEGL-3 (I)	Animal	3	3	1	1
Nitric acid	Significant	AEGL-2 (R)	Human	1	3	1	1
	Severe	AEGL-3 (R)	Animal	1	3	1	1
Phosgene	Significant	AEGL-2 (F)	Animal	3	3	1	1
	Severe	AEGL-3 (F)	Animal	3	3	1	1
Phosphine	Significant	AEGL-2 (R)	Animal	3	10[c]	1	1
	Severe	AEGL-3 (R)	Animal	3	10[c]	1	1
Sulfuric acid	Significant	ERPG-2	Animal/Human	NS	NS	NS	NS
	Severe	ERPG-3	Animal/Human	NS	NS	NS	NS

[a]Sparse data.
[b]The highest nonlethal value was close to the LC_{50} (concentration lethal to 50% of subjects) value.
[c]Children appear to be more susceptible.
Abbreviations: F, final value published by NRC (2000a, 2002, 2003); I, interim value under review by NRC; NS, not specified; P, proposed but not yet published in the *Federal Register* for public comment; R, interim value published in the *Federal Register* for public comment; U_H, factor that reflects uncertainty associated with interspecies variability (i.e., animal-to-human toxicity extrapolation); U_L, factor that reflects uncertainty associated with LOAEL-to-NOAEL extrapolation; U_D, modifying factor that reflects uncertainty associated with incomplete data; V, factor that reflects intraspecies variability in susceptibility to specified toxicity.
Sources: AIHA 1988, 1989, 1997; EPA 2002a,b,c,d,e,f, 2003a,b; NRC 2000a, 2002, 2003.

Cancer Assessments

Cancer assessment methods are currently in transition, so the methods used to assess cancer in TG-230 can only be understood by reviewing the underlying documentation. Generally, the cancer-risk values used by USA-CHPPM are based on older methodology that assumes that carcinogens have nonthreshold mechanisms and modes of action. Using that methodology, exposures to carcinogenic chemicals at any level are assumed to increase the risk of cancer development. Chemicals are classified into categories based on their likelihood of being carcinogenic to humans. For example, EPA (51 Fed. Reg. 33992 [1986]) used the following classifications:

- A. Known human carcinogen, based on adequate human data.
- B1. Probable human carcinogen, based on limited human evidence.
- B2. Probable human carcinogen, based on data from animals and inadequate or no evidence in humans.
- C. Possible human carcinogen.
- D. Not classifiable.

Newer methodologies (EPA 1999) use information on the modes and mechanisms of action of a chemical to assess risk, and a nonthreshold assumption is used only as a default when the mechanism of action is unknown. Agencies such as EPA evaluate the weight-of-evidence to estimate the degree to which each chemical might be a human carcinogen, and the assessment is described rather than categorized.

The typical quantitative cancer assessment underlying the Army's MEGs relies on linear extrapolation of the data from the concentrations in the study being used to zero. A "slope factor" is generated that is the upper bound (usually the 95% confidence limit) of the increased cancer risk from a lifetime exposure. Typically, that factor is expressed either as increased risk per unit dose (e.g., in units of risk per milligram per kilogram per day [$(mg/kg/day)^{-1}$]) or as increased risk per microgram per liter of drinking water ($[\mu g/L]^{-1}$) or microgram per cubic meter of air ($[\mu g/m^3]^{-1}$). The slope factor is expressed as the upper bound of risk. Thus, the risk is unlikely to exceed the upper bound value and is likely to lie somewhere between zero and the upper bound.

UFs are not applied in the traditional cancer assessment procedures. Newer methods that are more mechanism-based have options for considering linear and nonlinear extrapolations and the use of UFs. TG-230 and RD-230 should be updated to reflect the most recent approaches to cancer assessment, with the understanding that most of the existing cancer guidelines set by other agencies are not based on the newer methodologies.

Population Susceptibilities

Assumptions about sensitivity and susceptibility are incorporated into the development of health-related guidelines. Within an exposed population, exposures are rarely identical, and even when exposures are equal, doses to target tissues and cells are not the same in all people. People exposed to similar doses do not always have similar health outcomes. Protecting individuals or subpopulations that are more susceptible to adverse effects is a goal of most, if not all, health guidelines. The key question for deployed forces is who among the military population is likely to be more susceptible? Susceptible groups include those who might exhibit a greater effect in response to particular exposures. Some factors that might make individuals more susceptible include age, health status, and genetics. Historically, it has been assumed that the healthy men and women volunteers composing the military population would have few predisposing conditions that might make them sensitive or susceptible to environmental chemicals. In contrast, TG-230 and RD-230 assume that deployed populations include a more substantial representation of subpopulations that might be more sensitive to chemical exposures. USACHPPM's rationale for considering those subgroups in the development of MEGs is based almost entirely on a white paper by Weese provided in an appendix to RD-230.

According to data provided by USACHPPM (unpublished data, 2002) on the demographic characteristics of Persian Gulf War participants, the population in the theater of operations was 93% male with a median age of 24 years. The force consisted of 83% active duty personnel, and Army personnel made up 50% of the total. The crucial question about the deployed military population is whether it is different from the general population for risk-assessment purposes. Obviously, the two populations are not identical, but are there enough sensitive individuals in the military population to justify protecting for the same susceptibilities exhibited in the general population? TG-230 and RD-230 treat the factors in the military population that predispose individuals to sensitivity to chemicals as similar to and of the same magnitude as factors in the general population, excluding some groups such as children and the elderly. Some of the factors that were evaluated with regard to the military include genetic variability, asthma, and the embryo or fetus, primarily during the first trimester when pregnancy might not yet be detected.

Genetic variability is a subject that has received renewed interest due to the mapping of the human genome and recent research into genetic polymorphisms. Genetic variability is probably present in the military population at about the same level as it is in the general population. However, genetic variability does not comprise the whole of human variability in responses to chemicals, just as genetic make-up does not totally determine human responses in any other aspect of life. Variability in responses to chemicals results from differences in age, gender, nutritional status, and lifestyle factors in addition to genetic background (Calabrese and Gilbert 1993). Genetic variability that causes an individual to be sensitive to one type of chemical might not result in changes in sensitivity to other types of chemicals when the mechanisms of toxicity are different. The distribution of genetic variability might not reflect the ultimate variability in responses to a chemical, because additional compensatory mechanisms, such as redundant pathways, homeostatic mechanisms, and repair processes, could operate. At present, there is insufficient information to explicitly incorporate genetic susceptibility into exposure guidance except in the case of a few chemicals, including chemical warfare agents that act through cholinesterase inhibition.

There are about 17 million asthmatic individuals in the United States (AAFA 2003), and they make up about 6% of the general population. When reliably diagnosed at any age, asthma, including reactive airway disease, exercise-induced bronchospasm, or asthmatic bronchitis, is cause for rejection in appointments, enlistments, and inductions into the U.S. Armed Forces (U.S. Department of the Army 2002). Furthermore, if the asthma diagnosis is in doubt, tests for reversible airflow obstruction or airway hyperactivity must be performed prior to acceptance into the military. Asthma is a cause for referral to a medical evaluation board for possible separation from the service (U.S. Department of the Army 2002). Although complete physical examinations of service members are not conducted prior to deployment, medical records of possible deployment personnel are screened for medical conditions that would preclude the service members from duty. In the cases of service members who develop asthma while on active duty, the condition should be documented on their medical records. A service member with asthma can be placed on a temporary medical profile for 1 year. At the end of that year, he or she must be able to meet all the requirements for duty and training, including the running standards of the physical training test. It is expected that the percent of asthmatic individuals in the general military population and, more specifically, among deployed forces is much lower than that in the general population. Also, any cases of asthma that might be encountered during a deployment would be expected to be mild given the screening executed by the military.

According to the white paper by Weese presented in RD-230, only 15% of service members are female, and less than 7% of the forces deployed during the Persian Gulf War were female. Female soldiers are not deployed if they are known to be pregnant; however, there is the possibility that some women might not know they are pregnant at the time of deployment or might become pregnant during deployment. First trimester embryos or fetuses are considered a sensitive subpopulation by USACHPPM, even though the potential number of them in the deployed population is probably very small. However, some MEGs do not appear to be protective in this regard. For example, the 2-week water MEG for dinoseb is 0.14 mg/L, which would yield an exposure approximately 30 times greater than EPA's RfD of 0.001 mg/kg/day set for fetal protection (EPA 1989). Lead is also a special consideration (see Box 3-1).

Most of the existing guidelines used by USACHPPM to determine MEGs were developed for target populations other than the military. That means that different population characteristics and susceptibility factors were used in developing those guidelines. See Table 3-2 for the names of guidelines set by other organizations and the populations they were meant to protect. Guidelines developed for the general public usually include factors to protect susceptible subpopulations that might be more sensitive because of their age (e.g., infants, children, and elderly) or health status (e.g., pre-existing disease). Guidelines such as EPA's RfDs, RfCs, and acute exposure guideline levels (AEGLs) include UFs to calculate exposure values expected to be protective of those sensitive subpopulations. Other guidelines, such as the emergency response planning guidelines (ERPGs) and temporary emergency exposure limits (TEELs), are developed to protect most individuals in the general public but not particularly sensitive individuals. For guidelines in the workplace, such as Threshold Limit Values (TLVs) and immediately dangerous to life and health (IDLH) values, it is assumed that the worker population is composed of relatively healthy adults, and therefore, the standards are not designed to be protective of sensitive subpopulations. There are a few guidelines that have been developed specifically for the military population, including continuous exposure guidance levels (CEGLs) and field drinking water standards (FDWS). For those military standards, no special consideration was given to susceptible individuals. Table 3-3 shows the order of priority given to existing guidelines in establishing MEGs.

BOX 3-1 Evaluation of the Protectiveness of the 14-Day Water MEG for Lead

The water MEG for lead was calculated using a criteria value of 50 micrograms per liter (µg/L) published by the World Health Organization (1996).[a] Bowers (2003) modeled lead in blood at the request of the subcommittee using the specific parameters employed by USACHPPM. The model was derived from research by O'Flaherty (1993). The following exposure assumptions were made: a young woman born in 1980 ingested 15 L of water per day with lead at 50 µg/L for 2 weeks at the age of 23 and then returned to a normal water consumption rate of 2 L of water per day with lead at background concentrations. This results in a maximum blood lead concentration of 14 µg per decaliter (dL) for the woman, and it would typically take a little more than 6 months to return to original baseline levels. The embryo or fetus is generally assumed to have a blood lead level 0.9 times that of the mother, so fetal blood lead would peak at 12.6 µg/dL and would stay above 10 µg/dL, EPA's suggested maximum, for at least 2 weeks.

The protectiveness of the MEG is ambiguous. In fetal development, a day or even a few hours can mean the difference in susceptibility to the developmental toxicity of an environmental agent as well as the qualitative and quantitative nature of the effects. For that reason, exposures to the developing organism might be acute but have chronic outcomes. Thus, the number of days of exposure to the pregnant animal might be irrelevant. Therefore, an excursion of 26% over the accepted blood lead level might not be protective of the embryo or fetus, particularly in view of new findings that call into question the generally accepted "safe" level of 10 µg/dL (Canfield et al. 2003).

[a]The subcommittee subsequently discovered that WHO's drinking water guideline was incorrectly reported by USACHPPM. The correct value is 10 µg/L. Using the same exposure scenario above, but a lead concentration 10 µg/L, the peak blood lead level was found to be 4 µg/L. The blood lead level would return to baseline somewhat faster than the time estimated above for a 50 µg/L exposure. A 4 µg/L exposure would not be expected to affect the embryo or fetus, using the presently defined "safe" blood lead concentration of 10 µg/dL.

To consider susceptible subpopulations in the calculations for establishing noncancer exposure guidelines, a UF of 10 for human variability is typically applied (Haber et al. 2002). That factor is expected to cover differences in age, gender, genetics, and pre-existing disease that might make some individuals more susceptible to adverse effects from chemical exposures. Several sources have suggested that the UF of 10 should be divided into approximately equal pharmacokinetic and pharmacodynamic factors (Renwick and Lazarus 1998; Gentry et al. 2002). There is clear evidence of differences between the deployed military population and the general population that would tend to make the deployed military population less sensitive. For example, there are fewer asthmatic individuals (and presumably no severe asthmatic patients) in the deployed population than in the general population.

The assumption that the military population is as susceptible to the health effects of hazardous chemicals as the general population would lead to excessively conservative estimates of acceptable exposures. For example,

TABLE 3-2 Target Populations for Existing Exposure Guidelines Used in the Development of MEGs

Exposure Guideline	Organization	Target Population
AEGLs (acute exposure guideline levels)	U.S. Environmental Protection Agency	General public, including sensitive subpopulations
HAs (health advisories)	U.S. Environmental Protection Agency	General public, including sensitive subpopulations
MRLs (minimal risk levels)	Agency for Toxic Substances and Disease Registry	General public, including sensitive subpopulations
PMEGs (preliminary military air guidelines)	U.S. Army	General public, including sensitive subpopulations
RfDs (reference doses) RfCs (reference concentrations)	U.S. Environmental Protection Agency	General public, including sensitive subpopulations
SPEGLs (short-term public emergency guidance levels)	National Research Council	General public, including sensitive subpopulations
ERPGs (emergency response planning guidelines)	American Industrial Hygiene Association	General public
TEELs (temporary emergency exposure limits)	Department of Energy	General public
IDLH (immediately dangerous to life and health)	National Institute for Occupational Safety and Health	Worker population
TLVs (Threshold Limit Values)	American Conference of Governmental Industrial Hygienists	Worker population
EEGLs (emergency exposure guidance levels)	National Research Council	Military personnel
CEGLs (continuous exposure guidance levels)	National Research Council	Military personnel
FDWS (field drinking water standards)	U.S. Army	Military personnel

TABLE 3-3 Existing Guidelines Listed in Order of Priority for Use in the Development of MEGs

Short-Term				Long-Term		
Air						
1 hour	8 hours	14 days	Water	Air	Water	Soil
AEGLs[a]	AEGLs[a]	CEGLs	FDWS	PMEGs[a]	FDWS	RfDs[a]
ERPGs	TLVs	MRLs[a]	HA[a]	TLVs	HA[a]	
TEELs		TLVs	MRLs[a]	MRLs[a]	MRLs[a]	
Other		Special			RfDs[a]	
					Other	

[a]Protective of sensitive subpopulations.

MEGs for carbon monoxide were based on EPA's national ambient air quality standards (NAAQS), which were set to protect exercising angina patients. Angina patients would not be part of the deployed population, so that level of protection is not necessary for military personnel. A National Research Council (NRC 2000b) report makes the following important distinctions between military and civilian risk assessment:

> Incorporating "margins of safety" or conservative estimates of acceptable exposures, as is frequently done in environmental and occupational health settings, is not always useful to the needs of military risk management. When a high level of health and safety protection can be achieved without undue burdens or increases in other risks, such margins can be part of an effective risk-management program. But when risks must be borne or when probabilities of casualties must be weighed against immediate military considerations, best estimates of probable impact are more useful.

When the guideline used for deriving a MEG was designed to protect sensitive subpopulations (see Table 3-2), an adjustment factor should be applied (i.e., multiplied to remove the intraspecies UF that was used in the original derivation). For example, if an AEGL includes a UF of 10 to account for the susceptibilities of populations not likely to be deployed, that factor should be backed out of the guideline before it is used as the basis for a MEG. When the standard used to derive a MEG was designed to protect a military or healthy population (see Table 3-2), no UF should be applied to account for susceptible subpopulations unless there is a chemical- or population-specific reason to do so.

In determining MEGs, USACHPPM sometimes modified guidelines that incorporated an intraspecies UF by reducing the UF. For four short-term water MEGs (ammonium sulfamate, hexazinone, diisopropyl methylphosphonate, and isopropyl methylphosphonate) derived from EPAs health advisories (HAs), an adjustment was made to reduce the UF of 10 applied to protect for sensitive people in the general population. That was accomplished by multiplying the HAs by a factor of 3. USACHPPM considered applying a factor of 10 to all of the EPA's HAs, but decided to use the more conservative approach instead. The subcommittee is concerned that inconsistent adjustment of standards meant to protect the general population has resulted in inconsistently conservative MEGs (see Chapter 5 for discussion of media-specific MEGs for more details).

The subcommittee believes that military decision making would be better served by MEGs chosen consistently and likely to protect nearly all exposed deployed military personnel from chemical toxicity, consistent with DOD Safety and Occupational Health Program Instruction 6055.1 (August 19, 1998), to the extent feasible in the context of mission deployment. MEGs should reflect any well-documented differences in susceptibility or sensitivity to chemical-specific injuries between the military population and the general U.S. worker population. Such differences might be expected if the U.S. military deployed population has a demonstrated lower incidence of sickness, asthma, or obesity, or an absence of children or women past month 2 of pregnancy compared with the U.S. worker population. Any MEGs for deployed military personnel that are less restrictive than corresponding occupational guidelines used for U.S. workers (including DOD personnel stationed in the U.S.) should clearly be justified by reference to specific DOD-enforced operational conditions and/or to well-documented clinical-survey data.

In contrast, the subcommittee acknowledges that military decision making would be best served by CCEGs that employ no UFs for intraspecies differences unless there is specific evidence that some members of the deployed population are likely to be more sensitive to a specific chemical and that evidence would not otherwise be reflected in dose-response information used as the basis for a CCEG. That is, a UF for intraspecies differences should be applied to develop a CCEG only in cases where doing so improves the accuracy of that CCEG. For example, the accuracy of CCEGs would be improved in situations where the application of an intraspecies UF adjusts for bias introduced by using human data that inadequately reflect or do not reflect a sensitive subpopulation that reasonably can be anticipated to be present among deployed military personnel as the basis for dose-response estimation.

Exposure Factors

When assessing chemical risks, it is necessary to make assumptions about the rate of exposure through various routes. In TG-230 and RD-230, USACHPPM assumes that deployed troops have higher activity levels than the general population, which increases their ventilation and water consumption rates and thereby increases exposures to contaminants by those routes. USACHPPM evaluated available data on soldier-specific activities and exposure data from the EPA *Exposure Factors Handbook* (1990) and calculated a higher daily inhalation rate of 29.2 cubic meters (m^3) per day compared with EPA's default value of 20 m^3/day. Similarly, USACHPPM assumes that water consumption is much higher among deployed forces than the general public (which averages 2 liters [L] per day). Typically, the military assumes consumption of 5 L/day, but in dry, arid climates that rate could be as high as 15 L/day. These higher rates have been validated and established in Army doctrine (U.S. Department of the Army 1999) and are consistent with reports from the Israeli Defense Forces and U.S. Army Medical Services officers in the Mojave Desert (Henry 1985). The subcommittee supports the use of increased ventilation and water consumption rates for deployment risk assessments. It is important that these assumptions be consistently applied (see Chapter 5 for discussion of how the ventilation rate adjustment does not appear to have been applied to some of the 14-day MEGs).

Other exposure adjustments were sometime applied to the pre-existing guidelines to make them relevant to the exposure duration of interest to USACHPPM. Those adjustments are detailed and evaluated in Chapter 5.

ACCEPTABLE CANCER RISK

The Army requested that the subcommittee review its selection of an acceptable cancer risk of 1 in 10,000 (1×10^{-4}). The selection of an acceptable risk level is a policy decision, and the subcommittee does not believe it would be appropriate for it to make a judgment about how much risk the military should accept. However, the subcommittee decided that it could address this task by reviewing the acceptable risk levels selected by other organizations and making observations about where the Army's acceptable cancer risk threshold lies in comparison and the rationale used to set the threshold. With regard to chemical exposures, "safety" is often defined by various terms that include both scientific components and components that reflect societal values. In those instances where risk of injury can be quantified, safety also is expressed as levels of risk at which the consequences are considered to be of little or no concern (*de minimis* or acceptable risk) or require intervention (*de manifestis* risk).

The technical guides distinguish between chemical substances that can cause cancer and chemical substances that reportedly cannot. RD-230 indicates that health- or medical-risk acceptability applies solely to exposures to carcinogens, because noncarcinogens are governed by biological thresholds below which no injury is likely to occur. RD-230 further indicates that the acceptable risk of excess cancer resulting from exposures to chemical carcinogens is 1×10^{-4} regardless of route of exposure (e.g., inhalation, ingestion). Because 1-year MEGs are established using this risk level, the issue that is addressed is the incremental cancer risk averaged over a lifetime from a 1-year deployment. The rationales in RD-230 for the selection of the acceptable risk value are (1) that it is the upper bound of the range of cancer risk found acceptable to EPA (1×10^{-4} to 1×10^{-6}) (EPA 2001a) and (2) that it is an order of magnitude less than the acceptable level of risk generally supported for workers by the Occupational Safety and Health Administration (Rodricks et al. 1987).

The need to identify acceptable levels of risk rose to prominence in the debate over potential exposures to carcinogens present in the environment and in the workplace. On the basis of observations from radiation biology and theories of carcinogenesis, the concept that nonthreshold effects can result from exposures to carcinogens (or mutagens) was adopted for regulatory purposes.[1] The theory was that any exposure to a carcinogen carries with it some probability of an irreversible degree of damage, so no exposures to carcinogens can be judged risk-free, however small. In the past, pathological events thought to have a threshold were controlled by identifying the biological threshold, adjusting it by UFs, and keeping exposures below it (a "yes or no" decision). With the advent of the nonthreshold approach representing a continuum of risk, and the impracticability of maintaining a zero-risk policy, the issue of how much added risk was acceptable had to be addressed.

Identifying acceptable risk levels has been a subject of debate and disagreement for many years. For example, a Supreme Court decision regarding Section 112 of Clean Air Act cited a mandate that EPA identify "an acceptable level of risk" for human exposure to carcinogens (without regard to cost or technical feasibility) and to employ an "ample margin of safety"

---

[1]The concept that biological thresholds do not exist for carcinogens is no longer current (EPA 1999; see discussion earlier in this chapter). USACHPPM should update its technical guides accordingly.

to protect the public health (*Natural Resources Defense Council v. U.S. Environmental Protection Agency* 824 F.2nd1146 [1987])). In that case, however, the court was likely advocating, although indirectly, that a risk ceiling, above which inherently unsafe activities should be regulated without regard to cost, be used as the *de manifestis* acceptable risk level.

The court suggested that an acceptable risk level could be determined by adopting a "reasonable person standard." Risks associated with normal everyday activities (e.g., driving a car) and accepted by the general public could be considered acceptable (also referred to as "revealed preference"). Of course, that assumes that the true risks associated with activities are accurately known and understood by the participants in those activities. Given that the actual lifetime risk of a fatal car accident (2×10^{-2}), generally accepted by the public, is much higher than the estimated risks apparently tolerated for environmental pollution (1×10^{-4} to 1×10^{-7}) or occupational disease (1×10^{-3}), other factors clearly influence the decision of how much risk to tolerate, and the level of acceptable risk might well vary according to circumstance. As noted by Whipple (1988), the existence of a large risk does not excuse a small one when the benefits and other contextual factors are different, but social concerns and attention to risk should bear some relation to the magnitude of the risk under consideration. At this juncture, cost and benefit enter the decision-making process. There are examples of high-risk activities that have correspondingly high benefits being tolerated (e.g., some medical treatments); of high-risk activities that have low benefits being rejected; and of low-risk activities that have low benefits being evaluated on a case-by-case basis with considerable subjectivity.

Another means of selecting an acceptable risk level is to identify the risks associated with rare events that people face and presumably accept as consequences of everyday life (e.g., deaths from lightning strikes, tornadoes, bee stings, shark attacks). Actuarial data suggests those risks fall in the range of 1×10^{-6} (Whipple 1988).

Numerous scholarly works exist on the subject of acceptable risk (Lowrance 1976; Fischhoff et al. 1981; Whipple 1988). Most confine themselves to identifying the proper characteristics of the decision-making process for acceptable risk and the difficulties associated with that effort. Some of the issues used to judge acceptability include whether an activity is voluntary or involuntary; whether effects are immediate or delayed; whether alternatives are available; how well the risks are known; whether an item is essential or a luxury; whether the risk is encountered inside or outside of the workplace; whether the risk is common or especially dread; whether the average person is affected or only sensitive individuals; whether a product is used as intended or is likely to be misused; and whether the consequences are reversible or irreversible (Lowrance 1976; Slovic 1987).

The concept of acceptable risk, with its assorted nomenclature, has evolved and changed over the last few decades. The subcommittee presents a review in Appendix B that describes how various institutions and authors have defined acceptable risk as a concept applied to chemical products; it aims to distinguish between the scientific elements and value judgments.

Both RD-230 and TG-230 state that the Army's acceptable cancer risk is 1×10^{-4} and adjustments were made to some of the guidelines for carcinogens to ensure that the MEGs for carcinogens were based on the selected risk level. TG-230 indicates that the acceptable risk level is subject to change depending on the needs and characteristics of specific missions. The selected risk value for deployed military personnel falls within the range used by U.S. regulatory organizations and some international groups and is much lower than that applied to the nonmilitary workforce in the United States. This allows for some flexibility for situations involving repeated or multiple deployments where career-long exposure could lead to excess cancer risk greater than 1×10^{-4}. By establishing a 1-year guideline based on a target excess lifetime risk of 1×10^{-4}, it is reasonable to expect that career-long risk would not exceed 1×10^{-3}. Thus, 1×10^{-4} is consistent with DOD policy that "acceptable exposure measures and limits shall be derived from use of the risk management process" (DOD Instruction 6055.1, August 1998). The limit is reasonable in terms of protecting human health and also is flexible, allowing commanders to balance mission objectives and health protection.

One issue involving acceptable levels of risk bears discussion. Military documentation provides little usable guidance on variations in exposures where some of the exposure concentrations exceed the acceptable risk threshold for some length of time. Dealing with those fluctuations requires an understanding of the underlying toxicological and epidemiological information on which the risk estimates are based and of the parameters and outputs of the low-dose model or models employed.

CONSIDERATION OF IMMEDIATE AND LONG-TERM HEALTH EFFECTS

Army policy (U.S. Department of the Army 2001) dictates that a process be in place to

- "Ensure that commanders are aware of and consider the FHP-OEH [force health protection occupational and environmental health] risks created by OEH exposures (both long-term and short-term) during all phases of military operations, and over the broad spectrum of military activities."
- "Reduce the OEH exposures to as low as practicable to minimize short-term and long-term health effects in personnel within the context of the full spectrum of health and safety risks confronting the deployed personnel and consistent with operational risk management principles."

Thus, in reviewing the Army's technical guides, the subcommittee was asked to evaluate the balance of emphasis between health effects produced immediately or soon after chemical exposures and possible long-term or delayed health effects (e.g., cancer).

The need for considering immediate and long-term health effects is discussed in both TG-248 and TG-230 as part of the guidance on MEG development and implementation. The two types of health considerations for chemical risks are characterized as symptoms occurring either "during" or "after" the mission, as shown in Chapter 2, Table 2-2, which provides guidance on how to rank a chemical hazard's severity. Because military operational risk management (ORM) focuses on mission success, that guidance table was designed to categorize health effects that might occur during the mission and could affect the functional capabilities of personnel (i.e., medical threats) as of greater risk than delayed health effects.

The subcommittee found that for mission-risk assessment, it is generally necessary to focus the commander's attention on short-term effects. The guidance provided in TG-248 for occupational and environmental health and endemic disease (OEH/ED) hazard-severity ranking and TG-230's chemical hazard-ranking scheme appropriately give greater emphasis to short-term effects. In TG-248, post-mission symptoms are categorized as either having "negligible" or "marginal" effects on the mission, depending on the percentage of exposed personnel projected to exhibit the symptoms. For chemical exposures, TG-230 classifies post-mission symptoms resulting from exposures where MEGs (or CCEGs, as recommended by the subcommittee) either were not exceeded or were only minimally exceeded as *health* threats posing no threat or a negligible threat to the mission. However, the classification guidance given in TG-230 is limited to health outcomes that occur in 0-10% of personnel, clearly below any mission critical threshold identified in Tables 2-1 to 2-4. No explicit guidance is provided on how to classify situations involving delayed-onset or chronic illness that might occur in a larger percentage of the deployed population (e.g., chronic lung or liver disease that might result from acute exposure to lung or liver toxicants, respectively).

For the purposes of force health protection, short- and long-term health effects should be considered equally, because the goal is to protect the health of each individual that might be exposed. The example summary table provided in TG-230's Appendix F illustrates that USACHPPM's intent is that short- and long-term health information will be provided to decision makers simultaneously with mission-risk information. That intent should be made more explicit in the main body of TG-230.

Finally, the subcommittee noted that cancer is highlighted in the guidance as the most serious delayed outcome of chemical exposures, whereas other possible chronic or delayed maladies and dysfunctions are not as thoroughly addressed, if addressed at all. That implies that cancer is the only long-term consequence of concern to the military. To rectify this deficiency, attention should be given to other chronic or delayed-onset effects, such as effects on the respiratory system, the immune system, reproduction, development, and kidney function.

In the future, the subcommittee expects that some consideration should also be given to assessing indirect effects of exposure, such as psychological or morale effects, which could affect health and mission effectiveness. Because other aspects of deployment also cause or contribute to those stresses (e.g., hostile environment, separation from family), it might be appropriate that guidance be developed at the operational risk management level.

AGGREGATE EXPOSURE AND CUMULATIVE RISK

EPA defines aggregate exposure as exposure to a single chemical by multiple pathways (e.g., air, food, drinking water) and multiple routes of exposure (inhalation, oral, and dermal) (EPA 2001b). In TG-230, USACHPPM considers exposures from each environmental medium (air, water, and soil) independently and gives little consideration to aggregate exposures from multiple pathways. For the purposes of CCEGs, it is probably unnecessary to aggregate the risks from multiple media because air is likely to be the dominant source of exposure. However, it is important for force health protection that some consideration is given to aggregate exposures. (See Chapter 5 for further discussion.)

Cumulative risks from exposures to more than one chemical or to multiple hazards are also important considerations. Cumulative risk assessment involves studying the accumulation (over time, across sources, across routes, etc.) of stressors or exposures that can cause adverse effects and integrating the possible effects of those stressors to estimate and characterize the cumulative risk they pose (EPA 2001c). NRC (2000b) noted that "Troops during deployment could become exposed to a number of threats simultaneously. Exposures that are individually acceptable without appreciable risk might not be so when several are experienced together, and the question of interactions among agents looms particularly large for deployment risk assessment." TG-230 assumes that the toxicity of a mixture of chemicals that have similar modes of action will be equal to the sum of the weighted dose toxicities of the individual chemicals. Although that is generally an accepted practice, it is unclear how cumulative risk should be assessed when multiple hazards are present. RD-230 states that "TG-230 provides a general approach to address the potential for additive or even synergistic reactions when there are multiple chemical hazards present ... [as] exemplified through various Hypothetical Case Studies presented in 230 Appendix F in TG." Indeed, three of the seven hypothetical case studies (CS-3, CS-6, and CS-7) describe exposure scenarios that involve multiple chemicals, multiple potential routes of exposure, and/or multiple exposure locations. However, the potential difficulties in conducting comparative analyses of mixtures of threats are not illustrated by these cases studies, because each case described involves risks that clearly are dominated by a single risk source.

TG-230 does not provide guidance on how cumulative risks are to be assessed, other than that they should be considered qualitatively. The subcommittee examined a number of chemicals whose similar lethal effects would at least summate, but found that it was impossible to identify that type of potential interaction from the descriptions of symptoms and target organs provided in RD-230. For example, both hydrogen cyanide and hydrogen sulfide can cause death by terminally inhibiting oxidative metabolism, but the look-up tables do not indicate that potential.

For the purposes of mission-risk assessment, problems caused by the categorical analytic scheme are compounded by the lack of a systematic procedure to combine the multiple corresponding categorical levels of hazard severity and hazard probability that could be involved during operations. Any such procedure is likely to be too cumbersome to be practical and to yield results of questionable consistency—particularly in cases that involve multiple sources of relatively high risk from chemical exposures. An alternative procedure that could facilitate comparative analyses is discussed in Chapter 4 and in Appendix E.

For the purposes of force health protection, it is important to consider risks from exposures to multiple chemicals. This will require assessing, to the extent feasible, whether the toxicities of chemicals found in mixtures produce additive, less than additive, antagonistic, or synergistic effects. That might be accomplished in the short-term by flagging compounds that are likely to be present in mixtures and in combinations that might be of concern. Chemicals frequently affect the same organs by different mechanisms, but for the sake of simplicity, similar pathologies often are considered to share a common mechanism. Chapter 5 expands on the cumulative risk considerations for MEGs.

EXPOSURE ASSESSMENT

Reliable identification and assessment of chemical exposures is a key component of the application and interpretation of MEGs and CCEGs. Exposure assessment is the determination or estimation (qualitative or quantitative) of the magnitude, frequency, duration, and route of exposure to a particular chemical (57 Fed. Reg. 22888 [1992]). It involves the identification, measurement, and modeling of exposures to potential chemical hazards. Exposure assessment is discussed generally in TG-230 (particularly in Appendix F, "Hypothetical Case Studies") and in TG-248 (which expands upon the METT-TC considerations of mission, enemy, terrain, troops, time, and civilians), but no comprehensive guidance is provided on what exposure metrics (e.g., averaging times, peaks) the Army plans to use or how to develop and apply an exposure-assessment plan. The subcommittee was informed that more specific guidance is currently being developed by USACHPPM in a separate technical guide (*A Soldier's Guide to Environmental and Occupational Field Sampling for Military Deployment*).

For deployments, the subcommittee envisions that exposure assessments would, in general, involve identifying potential chemical hazards by using available classified and unclassified site-specific information; assessing the level of potential exposures by using sampling data, modeling, or assumptions; comparing exposure estimates with CCEGs to assess potential risks to the mission and to determine what risk trade-offs are necessary to accomplish the specific mission; and comparing exposure estimates with MEGs to assess potential health hazard and, in the event that some health trade-offs must be made, to determine what types of follow-up management actions are necessary to fulfill the military's force health protection responsibilities (e.g., documentation of exposures in medical records, medical monitoring).

Although it is clear that TG-230 was not designed to provide explicit guidance on how to conduct exposure assessments, the subcommittee believes it is worthwhile to go over some of the general requirements for an exposure-assessment plan. Such a plan should include, but should not be limited to, obtaining quantitative data from monitoring and modeling; developing a sampling plan; and establishing a decision logic that determines what actions to take based on the outcome of the exposure assessment (i.e., when to conduct additional sampling, when to require medical monitoring). The sampling plan should address the decision indicators for additional or reduced sampling frequency; the number and location of air, water, and soil samples; how to obtain representative and valid sampling results; and the need for continuous monitoring in some instances. Whenever relevant, decision-making aids, such as check lists or matrices, should be used. Statistical treatment and interpretation of data should be an integral component of the exposure-assessment guidance. For example, the large confidence limits associated with small data sets and the log-normal distributions typically found in occupational and environmental data sets increase the probability of misclassifications (i.e., concluding that exposures are acceptable when in fact they are unacceptable). Bayesian decision analysis is well suited to classifying data from limited data sets into one of several categories (e.g., clearly acceptable, marginally acceptable, marginally unacceptable, or clearly unacceptable) and could be used to develop decision logics for both force health protection and course-of-action decisions (Hewett 2003a,b). This technique requires limited quantitative exposure data in combination with simple professional judgment, previous analyses of historical exposure data, or exposure modeling predictions. The calculations are complex and require the use of programmable software; however, the user need only enter the exposure data and select an initial decision histogram. The end result is a final decision histogram of the probabilities that exposures occur in each of the exposure categories.

Army guidance should clarify the appropriateness of different exposure metrics for comparison with MEGs and CCEGs, and the differences in sampling methods, frequency, and intensity between exposure assessments conducted to support mission-risk assessments, those conducted to inform force health protection decisions, and others meant to provide documentation of personnel exposures to chemicals. Exposure assessments used for mission-risk assessment are particularly important, because time, access to external support, and data are more likely to be minimal in those situations. Identification and selection of appropriate risk-management techniques and the factors affecting the decision to proceed with their adoption should be included.

Force health protection exposure documentation might require sampling when there is little or no expectation of excess exposure. In other situations, several samples might be collected to verify worse-case exposures. Depending on the outcomes of exposure assessments, appropriate follow-up actions might include documentation of results with no further action, additional sampling, provision of personal protective equipment, substitution of materials or equipment, or cancellation of activity. Implementing sound industrial-hygiene practices that include documenting exposures and keeping records is an example of appropriate follow-up action.

Although some of the components of an exposure-assessment plan exist in the guidance, there is no logical, overall plan that indicates the philosophy, purpose, and principles of exposure assessment, particularly outlining the different approaches to exposure assessment for the purposes of force health protection and course-of-action decisions. If assembled, evaluated, and supplemented, the components already available would be valuable in developing a comprehensive exposure-assessment plan in circumstances that allow more thorough evaluations.

Situations might arise in which the limits of detection of available monitoring instruments are above the levels of concern for personnel exposures. Lists of the chemical agent detection technologies and the manufacturers of fielded instruments have been compiled in various documents (Brletich et al. 1995; IOM 1999; Jackson et al. 1998; Lewis and Lorenz 1998; O'Hern et al. 1997). The methods used for fixed-facility chemical warfare agent detection (mass spectrometry, gas chromatography, and Fourier transform infrared spectrometry) are not available in the field. The field technology currently available cannot provide the sensitivity and/or the rapid response necessary to protect troops from low concentrations of those agents. For example, for nerve agents GA, GB, GD, and VX, the minimal-effect air MEGs for 10 minutes to 24 hours are below the detection limits of handheld detectors (0.01 mg/m^3), as are the 24-hour MEGs that predict significant health effects.

Using exposure assessments to properly estimate risks requires a high degree of professional judgment on the part of preventive-medicine personnel, as noted in TG-230:

- "[Trained preventive-medicine personnel] should be familiar with basic methods of exposure assessment of chemicals in the environment. ... Military health services personnel will need to use professional judgment when applying the standardized information in this guide" (USACHPPM 2002a, p. 4).

- "The user should compare the guidelines with field sampling data or other (e.g., modeled) exposure data information. The interpretation of these comparisons will require professional judgment" (USACHPPM 2002a, p. 5).
- "The process of assessing and characterizing health risks from chemical exposures inherently involves significant data limitations, uncertainty, variability, and professional judgment" (USACHPPM 2002a, p. 19).

The level of professional judgment required for proper application of the MEGs and CCEGs makes it particularly important that preventive-medicine personnel receive comprehensive and adequate training in exposure assessment and risk management. The training should go beyond "how to" and "when and where" guidance; it should provide guidance on developing exposure-assessment plans consistent with the range of limiting conditions that are likely to be encountered and making risk-management decisions on the basis of exposure assessment outcomes. The latter issue is discussed further in Chapters 4 and 5.

UTILITY FOR DECISION MAKERS

The subcommittee was asked to evaluate the utility of TG-248 and TG-230 for decision makers, some of whom might not be knowledgeable about toxicology or the risk-assessment process. Because it was the subcommittee's understanding that the guides will be used in the field by a subordinate to the decision maker and not the decision maker, the subcommittee interpreted this task as asking if (1) an untrained individual could use the guides to characterize chemical risks appropriately and (2) if the risk characterization developed from using the guides would be useful to the decision maker. The first task seems to be in direct conflict with the statement in TG-230 that the guide is "not intended for use by untrained personnel or as a substitute for having trained preventive medicine personnel on-site or in the theater." TG-230 outlines an evaluative process that relies on the use of look-up tables, worksheets, and examples of how to apply those tools. The subcommittee found that TG-230 provides systematic guidance on how to evaluate potential chemical risks, but some of the decisions that must be made while using the guidance require the subjective judgment of experienced personnel. Therefore, some training in preventive medicine, toxicology, and risk assessment is necessary for TG-230 to be used effectively.

The second element of the question is whether the products of TG-230 guidance are easy for the decision makers to understand so they can properly consider the occupational and environmental health effects of chemical risks. The subcommittee concluded that the use of the ORM risk-assessment matrix in TG-230 effectively facilitates the communication of chemical-hazard risks in terms that are understandable to military decision makers. However, as noted in Chapter 2, TG-230 should be refined to make it consistent with TG-248, primarily to ensure that chemical risks are characterized using the same ORM metric as other risks. Chemical mission-risk estimates based on MEGs are not equivalent to other ORM risks because the MEGS are health-protection guidelines and not casualty estimates. CCEGs that predict casualty rates are necessary for appropriate course-of-action decision making.

RECOMMENDATIONS

This section summarizes the major recommendations made in this chapter on some of the key concepts and assumptions made in developing TG-248, TG-230, and RD-230. The text itself should be consulted for more thorough discussion of these issues and for several other recommendations that are of secondary importance.

- The application of UFs in setting exposure values to assess mission risks and health risks needs careful consideration. UFs should only be applied if they improve the accuracy of the exposure guideline for its intended purpose. Furthermore, thoughtful consideration will be needed to determine how to handle some of the complex issues involved in determining UFs, especially the UF for interspecies extrapolation, which involves pharmacokinetic and pharmacodynamic considerations.
- Immediate and long-term health effects should be considered when making course-of-action decisions. Greater weight should be given to immediate effects in the mission-risk assessment, and short- and long-term health consequences should be weighted equally in the health-risk assessment. USACHPPM needs to develop a more comprehensive set of guidance on how preventive-medicine personnel should convey long-term health information to commanders and what actions the Army should take to address those threats (e.g., when follow-up medical monitoring should be required).
- Deployed populations should be considered as healthier than the general population, and pre-existing health conditions do not need to be factored into the exposure guidelines. A UF for interindividual variation in susceptibility among humans should be applied to develop CCEGs only if

its application improves the accuracy of the CCEGs to predict toxic response by better accounting for evidence that a subpopulation within the deployed population is more sensitive than the group(s) used to obtain experimental or epidemiological data upon which that CCEG is based.

- TG-230 indicates that the embryo and fetus are of concern, so the current set of MEGs should be screened to assess whether they are protective of the embryo and fetus.
- Comprehensive exposure-assessment guidance should be compiled from existing sources and linked with TG-230 to support preventive-medicine personnel in developing exposure-assessment plans. The guidance should include information on monitoring and modeling, developing a sampling plan, and establishing a decision logic for management actions (discussed further in Chapter 4 and 5). The guidance should explain the differing approaches needed to support course-of-action decisions and to inform force health protection efforts.
- When CCEG and MEG values are below field detection capabilities, research and development support should be provided to aid in the development of more sensitive and reliable field detection equipment. This is particularly important in the case of chemical warfare agents.
- Guidance on how to consider risks from exposure to multiple chemicals, particularly in instances where there is no dominant risk source, is necessary. That will require assessment and documentation, to the extent feasible, of whether the toxicities of chemicals found in mixtures produce additive, less than additive, or synergistic effects.
- Because some of the decisions that must be made while using the guidance tools of TG-230 require subjective evaluation, it is important that personnel using the guidance have some training in preventive medicine and risk assessment.

REFERENCES

AAFA (Asthma and Allergy Foundation of America). 2003. Asthma and Allergy Foundation of America. [Online]. Available: http://www.aafa. org/

AIHA (American Industrial Hygiene Association). 1988. Formaldehyde Emergency Response Planning Guidelines. Fairfax, VA: AIHA Press.

AIHA (American Industrial Hygiene Association). 1989. Oleum, Sulfur Trioxide, and Sulfuric Acid Emergency Response Planning Guidelines. Fairfax, VA: AIHA Press.

AIHA (American Industrial Hygiene Association). 1997. Acrylonitrile Emergency Response Planning Guidelines. Fairfax, VA: AIHA Press. Brletich, N.R., M.J. Waters, G.W. Bowen, and M.F. Tracy. 1995. Worldwide Chemical Detection Equipment Handbook. Gunpowder Branch Aberdeen Proving Ground, MD: Chemical and Biological Defense Information Analysis Center.

Brletich, N.R., M.J. Waters, G.W. Bowen, and M.F. Tracy. 1995. Worldwide Chemical Detection Equipment Handbook. Gunpowder Branch Aberdeen Proving Ground, MD: Chemical and Biological Defense Information Analysis Center.

Calabrese, E.J., and C.E. Gilbert. 1993. Lack of total independence of uncertainty factors (UFs): Implications for the size of the total uncertainty factor. Regul. Toxicol. Pharmacol. 17(1):44-51.

Canfield, R.L., D.R. Henderson, Jr., D.A. Cory-Slechta, C. Cox, T.A. Jusko, and B.P. Lanphear. 2003. Intellectual impairment in children with blood lead concentrations below 10 microg per deciliter. N. Engl. J. Med. 348(16):1517-1526.

DOD (U.S. Department of Defense). 2003. Low-Level Chemical Warfare Agents (CWAs) Exposure Research Master Plan. June 2003.

EPA (U.S. Environmental Protection Agency). 1989. Dinoseb (CASRN 88-85-7), Oral Reference Dose Assessment. Integrated Risk Information System, U.S. Environmental Protection Agency. [Online]. Available: http://www.epa.gov/iris/subst/0047.htm [accessed May 2003].

EPA (U.S. Environmental Protection Agency). 1990. Exposure Factors Handbook. EPA/600/8-89/043. Exposure Assessment Group, Office of Health and Environmental Assessment, U.S. Environmental Protection Agency, Washington, DC.

EPA (U.S. Environmental Protection Agency). 2001a. Risk Assessment Guidance for Superfund: Vol. 1. Human Health Evaluation Manual (Part D, Standardized Planning, Reporting, and Review of Superfund Risk Assessments). Pub. 9285.7-47. Office of Emergency and Remedial Response, U.S. Environmental Protection Agency, Washington, DC. [Online]. Available: http://www.epa.gov/superfund/programs/risk/ragsd/front.pdf [accessed Dec. 2, 2003]

EPA (U.S. Environmental Protection Agency). 2001b. General Principles for Performing Aggregate Exposure and Risk Assessments. Office of Pesticide Programs, U.S. Environmental Protection Agency. [Online]. Available: http://www.epa.gov/pesticides/trac/science/aggregate.pdf [accessed Dec. 2, 2003].

EPA (U.S. Environmental Protection Agency). 2001c. Guidance on Cumulative Risk Assessment of Pesticide Chemicals That Have a Common Mechanism of Toxicity. Office of Pesticide Programs, U.S. Environmental Protection Agency, Washington, DC. [Online]. Available: http://www.epa.gov/pesticides/trac/science/cumulative_guidance.pdf [accessed Dec. 2, 2003].

EPA (U.S. Environmental Protection Agency). 2002a. Acrolein (CAS Reg. No. 107-02-8) Proposed Acute Exposure Guideline Levels (AEGLs). Proposed 1 Technical Support Document: 11/2002. Office of Pollution Prevention and Toxics, U.S. Environmental Protection Agency.

EPA (U.S. Environmental Protection Agency). 2002b. Bromine (CAS Reg. No. 7726-95-6) Proposed Acute Exposure Guideline Levels (AEGLs). Proposed 1 Technical Support Document: 05/2002. Office of Pollution Prevention and Toxics, U.S. Environmental Protection Agency.

EPA (U.S. Environmental Protection Agency). 2002c. Chlorine (CAS Reg. No. 7782-50-5) Interim Acute Exposure Guideline Levels (AEGLs). Interim 2 Technical Support Document: 12/2002. Office of Pollution Prevention and Toxics, U.S. Environmental Protection Agency.

EPA (U.S. Environmental Protection Agency). 2002d. Hydrogen Chloride (CAS Reg. No. 7647-01-0) Interim Acute Exposure Guideline Levels (AEGLs). Interim 4 Technical Support Document: 11/2002. Office of Pollution Prevention and Toxics, U.S. Environmental Protection Agency.

EPA (U.S. Environmental Protection Agency). 2002e. Hydrogen Sulfide, Interim Acute

Exposure Guideline Levels (AEGLs). Interim 4 Technical Support Document: 11/2002. Office of Pollution Prevention and Toxics, U.S. Environmental Protection Agency.

EPA (U.S. Environmental Protection Agency). 2002f. Acute Exposure Guideline Levels (AEGLs) for Phosphine (CAS Reg. No. 7803-51-2). Interim 1 Technical Support Document: 12/2002. Office of Pollution Prevention and Toxics, U.S. Environmental Protection Agency.

EPA (U.S. Environmental Protection Agency). 2003a. Hydrogen Fluoride (CAS Reg. No. 7664-39-3) Interim Acute Exposure Guideline Levels (AEGLs). Interim 4 Technical Support Document: 02/2003. Office of Pollution Prevention and Toxics, U.S. Environmental Protection Agency.

EPA (U.S. Environmental Protection Agency). 2003b. Acute Exposure Guideline Levels (AEGLs) for Nitric Acid (CAS Reg. No. 7697-37-2). Draft 2 Technical Support Document: 03/2003. Office of Pollution Prevention and Toxics, U.S. Environmental Protection Agency.

Fischhoff, B., S. Lichenstein, P. Slovic, S.L. Deby, and R.L. Keeney. 1981. Acceptable Risk. Cambridge: Cambridge University Press.

Gentry, P.R., C.E. Hack, L. Haber, A. Maier, and H.J. Clewell, III. 2002. An approach for the quantitative consideration of genetic polymorphism data in chemical risk assessment: Examples with warfarin and parathion. Toxicol. Sci. 70(1):120-139.

Haber, L.T., A. Maier, R.P. Gentry, H.J. Clewell, and M.L. Dourson. 2002. Genetic polymorphisms in assessing interindividual variability in delivered dose. Regul. Toxicol. Pharmacol. 35(2 Pt.1):177-197.

Henry, C.D. 1985. Heat stress and its effects on illness and injury rates. Mil. Med. 150(6):326-329.

IOM (Institute of Medicine). 1999. Chemical and Biological Terrorism: Research and Development to Improve Civilian Medical Response. Washington, DC: National Academy Press.

Jackson, W.M. et al. 1998. The Emergency Responders Ability to Detect Chemical Agents. Gunpowder Branch Aberdeen Proving Ground, MD: Chemical and Biological Defense Information Analysis Center.

Lewis, M., and M. Lorenz. 1998. Life Science Data in DOD Chemical and Biological Modeling and Simulation. ITT Industries, for Director, Environmental Life Sciences, Directorate, Defense Research and Engineering, 14 August 1998. (as cited in DOD 2003).

Lowrance, W.W. 1976. Of Acceptable Risk: Science and the Determination of Safety. Los Angeles, CA: W. Kauffman.

NRC (National Research Council). 2000a. Acute Exposure Guideline Levels for Selected Airborne Chemicals, Vol. 1. Washington, DC: National Academy Press.

NRC (National Research Council). 2000b. Strategies to Protect the Health of Deployed U.S. Forces: Analytical Framework for Assessing Risks. Washington, DC: National Academy Press.

NRC (National Research Council). 2002. Acute Exposure Guideline Levels for Selected Airborne Chemicals, Vol. 2. Washington, DC: The National Academies Press.

NRC (National Research Council). 2003. Acute Exposure Guideline Levels for Selected Airborne Chemicals, Vol. 3. Washington, DC: The National Academies Press.

O'Flaherty, E.J. 1998. A physiologically based kinetic model for lead in children and adults. Environ. Health Perspect. 106(Suppl. 6):1495-1503.

O'Hern, M.R., T.R. Dashiell, and M.F. Tracy. 1997. Chemical defense equipment. Pp. 377-382 in Medical Aspects of Chemical and Biological Warfare, Textbook of Military

Medicine, Part 1. Warfare, Weaponry, and the Casualty, Vol. 3, F.R. Sidell, E.T. Takafuji, and D.R. Franz, eds. Falls Church, VA: Office of the Surgeon General, U.S. Army.

Renwick, A.G., and N.R. Lazarus. 1998. Human variability and noncancer risk assessment —an analysis of the default uncertainty factor. Regul. Toxicol. Pharmacol. 27(1):3-20.

Rodricks, J.V., S.M. Brett, and G.C. Wrenn. 1987. Significant risk decisions in federal regulatory agencies. Regul. Toxicol. Pharmacol. 7(3):307-320.

Slovic, P. 1987. Perception of risk. Science 236:280-285.

USACHPPM (U.S. Army Center for Health Promotion and Preventive Medicine). 2001. Guide for Deployed Preventive Medicine Personnel on Health Risk Management. Technical Guide 248. U.S. Army Center for Health Promotion and Preventive Medicine. August 2001. [Online]. Available: http://chppm-www.apgea.army.mil/deployment/ [accessed November 25, 2003]

USACHPPM (U.S. Army Center for Health Promotion and Preventive Medicine). 2002a. Chemical Exposure Guidelines for Deployed Military Personnel. Technical Guide 230. U.S. Army Center for Health Promotion and Preventive Medicine. January 2002. [Online]. Available: http://chppm-www.apgea.army.mil/deployment/ [accessed November 25, 2003]

USACHPPM (U.S. Army Center for Health Promotion and Preventive Medicine). 2002b. Chemical Exposure Guidelines for Deployed Military Personnel. A Companion Document to USACHPPM Technical Guide (TG) 230 Chemical Exposure Guidelines for Deployed Military Personnel. Reference Document (RD) 230. U.S. Army Center for Health Promotion and Preventive Medicine January 2002. [Online]. Available: http://chppm-www.apgea.army.mil/deployment/ [accessed November 25, 2003]

U.S. Department of the Army. 1999. Sanitary Control and Surveillance of Field Water supplies. Draft (Technical Bulletin, Medical 577, Draft May 1999).

U.S. Department of the Army. 2001. Force Health Protection (FHP): Occupational and Environmental Health (OEH) Threats. HQDA Ltr 1-0-1. U.S. Department of the Army, Washington, DC. June 27, 2001.

U.S. Department of the Army. 2002. Medical Services, Standards of Medical Fitness. Army Regulation 40-501. Headquarters, Department of the Army, Washington, D.C. September 30, 2002.

Whipple, C. 1988. Acceptable risk. Pp. 157-170 in Carcinogen Risk Assessment, C.C. Travis, ed. New York, NY: Plenum Press.

WHO (World Health Organization). 1996. Guidelines for Drinking Water Quality, 2nd Ed. Geneva: World Health Organization.

4

A New Set of Exposure Guidelines: Chemical Casualty Estimating Guidelines

INTRODUCTION

As discussed in Chapter 2, predictive guidelines are needed to properly assess the chemical risks to missions. To that end, the subcommittee recommends that the Army develop and use chemical casualty estimating guidelines (CCEGs). CCEGs are media and duration-specific estimated chemical concentrations that would be expected to cause health impairments that degrade the performance of enough individuals to reduce unit strength (i.e., to pose a medical threat). They would be used to evaluate course-of-action options expected to involve chemical exposures. To be practical, CCEGs must accurately predict casualties and be in a form that can be applied in the field. This chapter addresses the development of CCEGs for individual chemicals, the application of CCEGs, and the estimation of cumulative risk. The development and application of MEGs is considered in Chapter 5.

The criteria in Box 4-1 indicate the differences between CCEGs and the U.S. Army Center for Health Promotion and Preventive Medicine (USA-CHPPM) military exposure guidelines (MEGs) and provide goals for CCEG development. The dichotomy presented has some apparent overlaps that will become more clear in practice. For example, a short-term exposure could reduce pulmonary function and could preclude healthy but suscepti-

BOX 4-1 Criteria for CCEGs and MEGs

CCEGs

- Include chemicals likely to be encountered in sufficient quantities to degrade mission effectiveness.
- Include health effects that are manifested within minutes, hours, or several days that could immediately affect the functioning of troops (e.g., loss of cognitive ability, loss of visual acuity, significantly reduced cardiopulmonary functioning, muscular weakness) performing a mission. Does not include long-term health effects (e.g., cancer).
- Consider exposure time frames of hours, days, and weeks, rather than months or years.
- Relate to the military population, which includes generally healthy adult men and women with typical variations in genetic susceptibilities.
- Provide exposure-response and population-response information, insofar as possible. Include concentrations likely to cause effects in humans, along with a description of the severity and incidence expected. This information would enable chemical threats to be weighed in comparison to other mission threats (e.g., Table 3-1 in TG-230 would be more useful).
- Provide guidance primarily for the air exposure pathway, because troops have no choice but to breathe the air (except when gas masks are used). Theoretically, the water pathway might influence CCEGs, but the availability of alternative sources of water makes it relatively less important. Water exposure scenarios of special concern should be identified and addressed. Soil also deserves some consideration, but is unlikely to be a significant source of exposure.

MEGs

- Include a large number of chemicals likely to be present in deployments.
- Include concerns over longer-term health of individuals recognize that exposures at these levels would have no to minimal impact on immediate missions.
- Include virtually all exposure durations from 1 hour to 1 year.
- Relate to the military population.
- Indicate protective levels (i.e., levels assumed to represent no adverse effects or very low risk) for the exposure durations of interest.
- Provide guidance for management actions when MEGs are exceeded.

ble soldiers from engaging in heavy exercise. At a higher concentration, the average soldier might be affected, further reducing unit strength. Long-term exposure at lower concentrations of the same chemical might not cause near-term effects on pulmonary function, but could cause alterations in lung structure much later. Some chemicals might produce a continuum of effects; others might elicit different acute and chronic effects. For example, the likelihood that short-term exposures to relatively low concentrations of chemicals would cause cancer many years later is remote. But, consider chemicals that adversely affect fertility. Knowledge that those chemicals are present at levels of concern would not affect the performance of troops during missions lasting only a few days, but would be useful in fulfilling the Army's responsibility to protect the long-term health of the force.

The goal is for command to have broad, reliable information for a full range of decision making. It is complicated by the variables included, but it begins with making maximum use of the data available to evaluate exposures that might affect imminent decisions or missions and exposures that are of longer-term concern.

DERIVATION OF CHEMICAL CASUALTY ESTIMATING GUIDELINES

The existing health-protective exposure guidelines set by other organizations usually do not satisfy the criteria for CCEGs outlined in Box 4-1. The closest in form are the U.S. Environmental Protection Agency (EPA) acute exposure guidelines levels (AEGLs) that are set for different exposure durations and increasing severity of health effects. However, AEGLs derivation includes consideration of susceptible subpopulations not present in the military population. Also, AEGLs typically are derived on the basis of a critical or most sensitive effect. In some cases, that particular effect might not influence mission-related performance. For example, the 1-hour MEG for nitrogen dioxide was developed to prevent "mild irritation" at the "severe effect level"; the 1-hour MEG for sulfur dioxide is based on alteration in pulmonary function in exercising asthmatic individuals; the 8-hour MEG for aldrin appears to be based on prolonged exposure leading to effects on the liver and central nervous system; and the 8-hour MEG for diesel fuel smoke is based on weight losses and focal pneumonitis in rats after multiple exposures. In addition, none of the 1- and 8-hour MEGs was adjusted with the military inhalation adjustment factor (see Chapter 3). Thus, the existing 1- and 8-hour MEGs are not directly useful to the quantitative assessments necessary for course-of-action risk comparisons. The subcommittee recommends that USACHPPM develop CCEGs by evaluating and using the primary scientific literature on individual chemicals as source material and deriving guidelines that meet the criteria in Box 4-1.

In this section, the subcommittee outlines approaches and concepts that should be considered in developing CCEGs. The discussion is intended as general guidance and not as detailed instructions on how to proceed, because alternative approaches are also possible.

Approaches for All Classes of Chemicals

The three major steps to developing CCEGs will be

- identifying the chemicals of interest,
- establishing standard procedures for developing CCEGs, and
- developing CCEGs for individual chemicals.

The subcommittee envisions that CCEGs will be necessary for a small subset of the chemicals for which MEGs have already been developed. Those would include chemicals that are likely to be encountered in sufficient concentrations to cause medical threats. They are likely to be air contaminants almost exclusively, because exposures to air contaminants are the most difficult to avoid or mitigate. However, the potential for water risks might need to be considered because of the variability in the quality of water sources available during missions. The subcommittee was told of an ongoing effort by an international task force (ITF-40) to prioritize acute chemical hazards and suggests that it might be a starting point for identifying chemicals of interest.

Standard procedures should be developed to derive CCEGs for individual chemicals. That will help ensure that the CCEGs specifically address the needs of the U.S. Department of Defense (DOD), are developed on a consistent basis, and are distinguishable from other standards used within the military (e.g., MEGs and standards from the Occupational Health and Safety Administration). Similar procedures have been developed by other organizations for similar purposes, including the standing operating procedures for developing EPA's acute exposure guideline levels (AEGLs) (NRC 2001) and the National Aeronautics and Space Administration guidelines for developing spacecraft maximum allowable concentrations (NRC 1992). Some of the considerations that should figure in the methodology for deriving CCEGs are discussed below.

Making appropriate comparisons of estimated impacts on troop viability and vulnerability requires two types of information: (1) the severity of the immediate medical consequences during the course of the mission, and (2) the likely number of troops affected in the exposure scenario envisioned for the specific mission. In addition, the commander should have information on potential long-term health effects, which might be captured best from MEGs. Although that information is not relevant to mission performance, it is essential to overall force health protection. Knowledge of potential long-term hazards might enable commanders to avoid those hazards without significant disruption to the mission.

The Army should carefully identify the relevant end points for casualty estimation, because those end points likely will be different from those used to establish health-protective exposures guidelines. For health-based standards, the adverse effect occurring at the lowest dose level in experimental studies is selected as the critical effect, under the assumption that an exposure level based on that effect will provide protection from other effects that occur at higher doses. The goal of CCEGs is to provide risk estimates of impacts on troop strength, including consideration of individuals affected to different extents by the exposure. Because there is a spectrum of effects that could have an impact, depending on the specific mission, it would be useful to consider categorizing health effects by graded severity levels, such as mild pathological responses (e.g., sensory discomfort, irritation, mild nonsensory effects), moderate pathological responses (e.g., temporarily debilitating systemic dysfunctions), and severe pathological responses (e.g., reversible or irreversible damages to organ functions that are incapacitating, life-threatening, or lethal). That scheme resembles the graded AEGLs in several respects.

The data on individual chemicals should be subjected to some form of weight-of-evidence analysis in which the quality of the data is examined critically and the degrees of consistency and concordance are evaluated closely. The process should include some rules for deciding the relative value of, and reliance on, human data versus animal data. In addition, many of the studies relevant to CCEGs will not have been designed for such a purpose. Thus, confidence in the incidence level per unit of dose and dose range can range from high to low depending on how well the range has been bracketed by field observations or experimental studies. The challenge for the military will be in not only defining the uncertainty for each data set but also in assuring consistency in the application and interpretation process.

CCEG values must be unbiased estimates of risk. They should be predictive estimates of casualties and they should not incorporate margins of safety or adjustments for missing information except under unusual circumstances. The available data should be used to derive the CCEGs and to inform the selection of uncertainty factors (UFs). For example, when the CCEGs are based on animal studies, interspecies extrapolation, typically assumed to consist of pharmacokinetic and pharmacodynamic components, is required. In some cases, data will be available to estimate pharmacokinetic differences or to at least allow for allometric adjustments. However, in most cases, the database will be insufficient to quantitatively assess pharmacodynamic differences, and it might be necessary to use a UF. The need to consider pharmacokinetic, pharmacodynamic, and other factors in selecting an interspecies UF has been described by the NRC (2001).

The data supporting some of the CCEGs could have significant deficiencies. For example, mortality data might be drawn from accidental exposures in a small number of people, raising concerns about the data's reliability to military populations and missions. If assessment and evaluation suggests that the data analyses could be biased by database deficiencies, an adjustment factor of 3 might be warranted.

As predictive values, the CCEGs should not include protection for civilian population sensitivities (e.g., pre-existing disease), which is typically achieved by applying an intraspecies UF. When chemicals of interest have known susceptible subpopulations, it might be necessary to formulate more reliable estimates of the mean responses of the entire deployed population at risk if those susceptible groups were not represented appropriately in the key studies. The CCEGs should consider factors that might increase doses in deployed personnel (e.g., higher ventilation rates, greater water consumption), as discussed in Chapter 3.

CCEGs should be designed to inform decision making within a relatively short time frame. The subcommittee assumes that most missions requiring course-of-action evaluation would be short (i.e., less than a few days). Although a mission might be executed over several days, specific events requiring CCEG evaluation would be more brief (i.e., less than 24 hours). If different time frames were of interest, the duration adjustment would be based on $C^n \times t$, and n would be determined by the information available and the slope of the dose-response curve.

In Appendix C, the subcommittee presents an illustration of one of the possible approaches to creating CCEGs, namely probit analysis. The appendix shows that knowing the percentage of troops responding (at three severity categories—mild, moderate, and severe) to a variety of exposure scenarios would be useful to commanders when comparing the possible risks to mission success. However, the subcommittee's illustration also shows how limitations in the underlying database could preclude the use of probit analyses for some chemicals of interest. Furthermore, the Army must consider how to factor in the issue that at levels at which casualties are seen, there will be a fraction of the exposed population that is affected to a lesser degree and that those lesser effects could also degrade mission capability.

The methodology for developing the CCEGs as well as the draft CCEG values should be externally peer-reviewed before their application. Because the CCEGs will be based on advanced quantitative analysis of available data and theory, they will incorporate a lot of scientific judgment. In addition, they will be predicting casualties, not suggesting safe levels buffered for error. Hence, peer-review will be especially useful to assess the robustness of the CCEGs.

The subcommittee recognizes the difficulty of this effort and the time needed to address it. One approach to minimizing the workload might be to form working relationships with organizations currently developing acute health guidelines.

The subcommittee was told that TG-230 is being used in the field. Until CCEGs are developed, it is important that USACHPPM amend TG-230 to warn users of the guide that MEGs should be applied with some caution. Because they are protective in nature, it is exceptionally difficult (if not impossible in some instances) to use them for direct comparisons with other operational threats.

Chemical Warfare Agents

CCEGs for chemical warfare agents (CWAs) might be needed in some circumstances, because exposures to those chemicals are expected to be acute; the effects can be severe; the potential for their use has several parallels with other operational threats; and troops will likely have personal protective equipment to greatly reduce the chances of exposure. This section reviews how USACHPPM developed MEGs for CWAs, illustrates some of the special considerations for those agents, and discusses the difficulties of using existing exposure guidelines for those chemicals as the basis for CCEGs, particularly with regard to the use of UFs.

In RD-230, AEGL values for CWAs are used directly as air MEGs for the 1-hour and 8-hour exposure durations. The 24-hour air MEGs for the agents were derived by straight-line extrapolation of the 8-hour AEGLs ($C \times t = k$). According to RD-230, analysis of CWA exposure scenarios indicates that a continuous exposure of deployed personnel to nerve agents or vesicants for a time period greater than 24 hours is very unlikely. Therefore, there are no MEGs for the agents for time periods greater than 1 day. USACHPPM did not use other toxicity estimates, such as the Army's acute human toxicity estimates (NRC 1997), for developing air MEGs. Those values were derived for wartime operations and casualty estimation on a gross scale.

RD-230 asserts that although AEGLs are designed for the general population, the AEGLs for CWAs are not overly conservative for military personnel on the basis of following arguments:

• For nerve agents, the identified susceptible subpopulations are those with abnormally low cholinesterase activity, which is a genetic sensitivity and is not screened out in the military.

• For sulfur mustard, the key health concern is effects on the eye. The variation in susceptibility to those effects in the military population is similar to that in the general civilian population.

• In general, variations in susceptibility among military personnel and variations within the general population are believed to be similar.

AEGLs are developed for the general population and usually consider individuals that might be more sensitive because of their age (e.g., infants, children, and the elderly) or health status (e.g., the ill or infirm) in their calculations. In the case of CWAs, one or more UFs were incorporated. Appendix D is a summary of the critical studies and UFs used in the development of the AEGL-1, AEGL-2, and AEGL-3 values for GB (adapted from NRC 2003). The AEGLs for GA, GD, GF, and VX were developed using GB data and a relative potency approach.

A default UF of 10 was used for intraspecies variability (protection of susceptible populations) in the development of the AEGLs for all G agents and VX. The differences among individuals in (1) blood cholinesterase and carboxylesterase activity; (2) gender (female subjects being more sensitive); (3) polymorphic paraoxonase gene (PON1); (4) levels of paraoxonase (which are particularly low in newborns); and (5) age-related sensitivity were discussed as sources of intraspecies variability in the NRC (2003) report. Given the criterion that CCEGs relate to the military population, a UF of 10 for intraspecies variability might overstate the expected adverse outcome, particularly when the exposure estimates are based on female rat data used to derive the AEGL-1 and AEGL-3 for GB (see Appendix D). Thus, the application of that UF should be re-examined in CCEG development.

The subcommittee agrees with USACHPPM that the NRC's recent assessment of CWAs for the development of AEGLs (NRC 2003) provides the most comprehensive evaluation of the existing data and should be relied on in the development of CCEGs. However, the subcommittee believes that the direct use of AEGLs as CCEGs would be inappropriate, because that approach does not satisfy the criterion that CCEGs provide exposure-response and population-response data, including the concentrations likely to cause effects in humans, along with descriptions of the magnitude and incidence of the expected effects. However, the supporting studies and end points for AEGL-1 and AEGL-2 should provide that information (see Appendix C for example using GB).

Other issues to consider when setting exposure guidelines for CWAs include the following:

- The $C \times t = k$ temporal extrapolation from 8 hours to 24 hours by USACHPMM should be re-examined in light of the $C^n \times t = k$ relationship described in the AEGLs documents.
- The NRC (2003) AEGL report indicates that accounting for breathing rates is not necessary for local effects (e.g., miosis) but is necessary for systemic effects. However, it did not include a factor for breathing rates in the development of AEGLs because of numerous uncertainties associated with the issue. The ERDEC-TR-489 provides occupational exposure limits for G agents that incorporate adjustments from experimental breathing rates to occupational conditions. The air MEGs for toxic industrial chemicals were developed on the basis of the estimated breathing rates (29.2 m^3/day) in soldiers. Differences between experimental breathing rates (under various conditions and different species) and actual battlefield breathing rates should be considered in the development of CCEGs for CWAs.
- An intraspecies UF of 3 was incorporated into the AEGLs for sulfur mustard to protect potentially sensitive individuals (Appendix D). However, the NRC (2003) report notes that there was little variability in ocular responses among subjects (Anderson 1942). For purposes of CCEGs, that UF introduces an unnecessary level of conservatism. A modifying factor (MF) of 3 was used in the AEGL-2 to account for the potential onset of long-term ocular and respiratory effects. That also is unnecessary for the CCEGs. The AEGLs for sulfur mustard, like the AEGLs for G agents, should not be used directly to develop CCEGs. The source data (critical study or studies) should be used to develop CCEGs in accordance with the criteria described in Box 4-1.

APPLICATION AND INTERPRETATION OF CCEGs

The CCEGs, which ultimately are intended to be used in the field, need to be readily interpretable by those who will apply them. A substantial amount of highly technical material must be evaluated comprehensively and synthesized into a summary matrix that can be used by preventive-medicine personnel in the field. The effort will require expressing probabilistic CCEGs, as illustrated in Appendix C, in the framework of the operational risk-management (ORM) risk levels to enable simultaneous evaluation and comparison of all the main classes of operational risks. The subcommittee recommends that a table be created for each CCEG chemical defining the concentrations that correspond with the boundaries between green and amber, amber and red, and red and black unit-strength levels (i.e., between low and moderate, moderate and high, and high and extremely-high risk levels) as defined in TG-230.

Table 4-1 provides an example of how CCEGs derived in Appendix C for seven chemicals could be used to estimate impacts on troop strength in place of the three-step categorical MEG-based procedure now described in TG-230. The chemical concentrations estimated to severely affect 15%, 30%, 40%, and 50% of the unit are listed with the corresponding ORM risk level and unit status, assuming that the entire unit is exposed. Any measured or modeled field concentration of any given chemical can be compared with the values in the table to estimate the potential impact on the mission in the color-coded ORM terms used by the decision maker.

To make this comparison, a measured or modeled concentration (C) for a given chemical would be compared with each listed concentration (C_{test}), starting with the right-most column and proceeding left until the first instance in which $C \geq C_{test}$; the color of that current table column then defines the unit status.

Table 4-1 can always be used to determine ORM risk levels by comparing measured or modeled concentrations with table entries when 100% of the unit is assumed to be exposed (as explained in footnote *a*). However, in cases where only a percentage of the unit (i.e., a subunit) is exposed and C is greater than the concentration that corresponds to a "green" unit status, an additional calculation might be required to derive the percentage of the entire unit affected. That calculation determines the corresponding ORM risk level. The percentage P^* of a unit affected seriously or severely (i.e., in a mission-incapacitating way) by chemical exposure is defined as $P^* = P \times F$, where F is the estimated fraction of the unit exposed to the chemical and P is the predicted percentage of that fraction that will be severely affected by the exposure. It is P^*, not P, that must be used to determine the color-coded ORM risk level, because those levels are defined in terms of troop strength percentage ranges (i.e., 100% - P^*). The following procedure can always be used to calculate P^* when <100% of the unit has been exposed, using only information in a CCEG table such as Table 4-1.

In the context of a mission-threatening (i.e., severe) toxic response from acute respiratory exposure to a specified chemical, the lognormal dose-response model that was used to generate Table 4-1 predicts that the percent P of exposed individuals to incur a specified toxic response is lognormally distributed as a function of ambient concentration C (i.e., as the quasi-threshold lognormal exposure-response function).

TABLE 4-1 Sample CCEGs for Seven Chemicals for "Severe" Response[a]

	Approximate Concentration in Breathing Zone (ppm-hour)			
Chemical	C_{15}	C_{30}	C_{40}	C_{50}
Aniline	1,400	1,600	1,800	1,850
1,1-Dimethylhydrazine	250	540	800	1,400
Hydrogen sulfide	640	680	700	710
Hydrogen cyanide	95	115	130	140
Propylene glycol dinitrate	40	70	90	120
Acrolein	34	40	45	48
Sarin	10	30	52	90
Evaluation	**Degree of Medical Threat**			
% of unit severely affected, P^*	15%	30%	40%	50%
Unit troop strength[a]	85%	70%	60%	50%
ORM risk level[a]	Low	Moderate	High	Extremely High
Unit status[a]	Green	Amber	Red	Black

[a]Assumes 100% of the unit is exposed.

Abbreviations: C_{15}, concentration estimated to effect 15% of the unit; C_{30}, concentration estimated to effect 30% of the unit; C_{40}, concentration estimated to effect 40% of the unit; C_{50}, concentration estimated to effect 50% of the unit.

Unit Status
Black: Unit requires reconstitution. Unit below 50% strength.
Red: Combat ineffective. Unit at 50-69% strength.
Amber: Mission capable, with minor deficiencies. Unit at 70-84% strength.
Green: Mission capable. Unit at 85% strength or better.

$$P = \Phi[\sigma^{-1}\log(C / C_{50})] \times 100\% \quad (4\text{-}1)$$

where F is the cumulative normal (Gaussian) probability distribution function, log denotes logarithm (using any specified base, such as 10 or e), C_{50} is the model ("location") parameter that represents the concentration that elicits a 50% response, and s is the model ("shape") parameter the inverse of which specifies the steepness of the dose-response curve. This lognormal model has only two estimated parameters, C_{50} and s. The parameter s may be defined in terms of the concentrations C_{15} and C_{50} (defined in Table 4-1 above) as $s = 0.9648 \log(C_{50}/C_{15})$. Therefore, Equation 4-1 can be rewritten as the following function of the measured (or modeled) concentration C and the concentrations C_{15} and C_{50} obtained from Table 4-1:

$$P = \Phi\left[1.036\frac{\log(C / C_{50})}{\log(C_{50} / C_{15})}\right] \times 100\% \qquad (4\text{-}2)$$

After first calculating P using Equation 4-2, the percentage $P^* = P \times F$ can be calculated as described above to estimate the unit status.

AGGREGATE EXPOSURE AND CUMULATIVE RISK

As discussed earlier, air is the most likely exposure pathway of concern for CCEGs. Although aggregate exposures theoretically could occur and could affect mission performance, such a scenario is unlikely. Therefore, establishing formal procedures for CCEGs for aggregate exposures is not a high priority. However, risks are likely to accumulate, making procedures to assess medical risks from mixtures of chemicals desirable. As it did for individual chemicals, the subcommittee recommends a well-grounded and applicable approach for mixtures.

Analytic frameworks that have occurrence probabilities modeled explicitly as functions of corresponding chemical exposure have been developed specifically for application to quantitative health-risk assessments involving multiple toxic chemicals and/or multiple toxicity end points (NRC 1994; Bogen 2001). This framework can be generalized to illustrate a quantitative approach that could be used to estimate the percent P^* of a unit expected to be affected by mission-hindering (typically serious) symptoms resulting from acute respiratory exposure(s) to multiple chemicals that might affect similar toxic end points and/or different toxic end points. Using that approach, P^* is calculated by applying a basic rule for aggregating independent likelihoods (known as de Morgan's rule [Parzen 1960; Ang and Tang 1975]) to an appropriate modification of Equation 4-2, above. The quantitative approach recommended by the subcommittee for exposures to multiple chemicals is described in Appendix E.

RECOMMENDATIONS

The development of CCEGs for informing course-of-action decisions in the field is of highest importance. To assist in obtaining and managing resources for that effort, DOD should analyze the staff and funding necessary to accomplish the subcommittee's recommendations and should establish priorities and a time estimate for the work. This section summarizes the major recommendations made in this chapter for the development of CCEGs. The text itself should be consulted for more thorough discussion of these issues and for several other recommendations that are more specialized, are more chemical-specific, or are of secondary importance.

- **CCEGs should be developed and used in TG-230 to support mission risk assessment.** This will involve identifying the set of chemicals for which CCEGs should be created, developing a method for creating them, and performing the necessary analyses for each chemical. The following are important elements to consider in addressing this recommendation:

—CCEGs should be established for those chemicals having some finite probability of being encountered in sufficient quantity to degrade unit effectiveness.

—CCEGs should be derived primarily for air contaminants, because inhalation is the exposure route most likely to result in incapacitation. The potential for water risks should also be considered, depending on the quality of the water supplies during particular missions. Some consideration might also be given to the need for assessing risk from less conventional routes of exposure, such as water emersion, that might occur with small unit and special operations.

—CCEGs should provide predictive, probabilistic exposure-response information that will enable chemical threats to be weighed in comparison with other mission threats. CCEGs ideally would be determined by modeling chemical-specific data to predict effects on unit strength at various exposure levels (e.g., probit analysis).

—The methodology for deriving CCEGs and the derivation of the CCEGs themselves should be peer-reviewed.

—Assistance from other organizations working on health-related guidelines should be pursued. Many existing exposure guidelines (especially EPA's AEGLs) have key information available in their documentation. Future working relationships between DOD and other agencies that routinely develop acute exposure guidelines might make the development of CCEGs more resource-effective.

—If the Army chooses to use MEGs in the interim, TG-230 should be revised to clearly articulate the deficiencies of MEGs and their limitations for assessing mission-related performance risks.

- **CCEGs should be provided to users in an ORM format different from that currently proposed in TG-230.** Personnel in the field will need to be able to make comparisons rapidly between estimated exposure concentrations for specified durations and the CCEGs to identify estimated unit status (i.e., green, amber, red, black) and to weigh chemical threats against other operational risks.
- **Consider cumulative risks in the ORM context.** The number of chemicals for which CCEGs are needed appears limited, making it feasible to identify those compounds that are likely to be present in mixtures and combinations of concern. Additivity assumptions could be applied using a probabilistic method consistent with the probabilistic nature of the CCEGs.

REFERENCES

Anderson, J.S. 1942. The effect of mustard gas vapour on eyes under Indian hot weather conditions. CDRE Report No. 241. Chemical Defense Research Establishment (India).

Ang, A.H-S., and W.H. Tang. 1975. Probability Concepts in Engineering Planning and Design, Volume I: Basic Principles. New York, NY: John Wiley & Sons. pp. 34-35.

Bogen, KT. 2001. Methods for Addressing Uncertainty and Variability to Characterize Potential Health Risk from Trichloroethylene Contaminated Ground Water at Beale Air Force Base in California: Integration of Uncertainty and Variability in Pharmacokinetics and Dose-Response. UCRL-ID-135978 Rev. 1 (www.osti.gov/servlets/purl/793701-BslGOu/native/). Lawrence Livermore National Laboratory, Livermore, CA.

EPA (U.S. Environmental Protection Agency). 2002. A review of the reference dose and reference concentration processes. EPA/630/P-02/002F.

EPA (U.S. Environmental Protection Agency). 1994. Methods for derivation of inhalation reference concentrations and application of inhalation dosimetry. EPA/600/8-90/066F. Available from NTIS Springfield, VA.

NRC (National Research Council). 2003. Acute Exposure Guideline Levels for Selected Airborne Chemicals. Volume 2. Washington, DC: The National Academies Press.

NRC (National Research Council). 2001. Standard Operating Procedures for Developing Acute Exposure Guideline Levels for Hazardous Chemicals. Washington, DC: National Academy Press.

NRC (National Research Council). 1997. Review of Acute Human-Toxicity Estimates for Selected Chemical-Warfare Agents. Washington, DC: National Academy Press.

NRC (National Research Council). 1994. Science and Judgment in Risk Assessment. Washington, DC: National Academy Press.

NRC (National Research Council). 1992. Guidelines for Developing Spacecraft Maximum Allowable Concentrations for Space Station Contaminants. Washington, DC: National Academy Press.

Parzen, E. 1960. Modern Probability Theory and Its Applications. New York, NY: John Wiley & Sons. pp. 11-16.

5

Process for Establishing and Applying Military Exposure Guidelines

The health-protective nature of the current military exposure guidelines (MEGs) makes them most appropriate for use as part of the Army's force health protection initiative. In this chapter, the subcommittee reviews how MEGs were derived in Reference Document 230 (RD-230) (USACHPPM 2002) and provides comments and recommendations on their application. In addition, the subcommittee considers the need to address risks from multiple exposure pathways, multiple chemicals, and repeated deployments.

AIR EXPOSURE GUIDELINES

This section provides a summary of the U.S. Army Center for Health Promotion and Preventive Medicine's (USACHPPM's) approaches to deriving its current air MEGs. The basic approach to determining air MEGs was to review the exposure guidelines of other agencies, use a hierarchical scheme to select the most appropriate guideline for the exposure duration of interest, and then adjust that guideline to meet the military's needs, if necessary. Air MEGs were developed for exposures of 1 hour, 8 hours, 14 days, and 1 year. In the sections below, the process used to derive the duration-specific MEGs is described and evaluated, followed by a review of some chemical-specific MEGs and criteria air pollutants.

1-Hour MEGs

Derivation

One-hour air MEGs were developed to consider three levels of health effects:

- **Minimal effects.** Above this level, individuals could begin to experience mild, transient effects that should not impair performance.
- **Significant effects.** Above this level, individuals could begin to experience irreversible or serious effects that might degrade performance and incapacitate a small portion of the people exposed.
- **Severe effects.** Above this level, some within an exposed population could begin to experience life-threatening or lethal effects.

The hierarchy used to select source material was (1) acute exposure guideline levels (AEGLs), (2) emergency response planning guidelines (ERPGs), (3) temporary emergency exposure limits (TEELs), and (4) other.

AEGLs are developed by the U.S. Environmental Protection Agency (EPA) and reviewed by a National Advisory Committee and by the National Research Council (NRC). AEGLs are developed for three severity levels, and all of the values are intended to protect the general public, including sensitive and susceptible subpopulations. Above AEGL-1 concentrations, the general population could experience discomfort and irritation effects that are not disabling and are reversible upon cessation of exposure. Above AEGL-2 concentrations, the general population could experience irreversible or serious health effects or impaired ability to escape. Above AEGL-3, the general population could experience life-threatening health effects or death. These levels were developed for exposure durations of 10 minutes (min), 30 min, 1 hour, 4 hours, and 8 hours. All AEGL-1 and AEGL-2 exposures were reviewed by EPA to ensure that they do not pose an excess cancer risk greater than 1×10^{-4}.

ERPGs are developed by the American Industrial Hygiene Association (AIHA) and are intended for emergency planning and response operations. They also have three levels of health effects that are quite similar to those of the AEGLs. They were created to target the general population, but not particularly susceptible individuals. TEELs are developed by the U.S. Department of Energy and are essentially interim ERPGs.

Evaluation

The hierarchy for selecting sources for the 1-hour MEGs is based on a logical argument and is consistent with the NRC (2000) recommendations for developing standards. The NRC (2000) noted that in the development of guidelines, different kinds of guidelines are appropriate for different settings. The report also stated that it is useful to allow for guidelines that permit some degree of toxic response but protect against incapacitation or irreversible injury for use in decision making during emergencies or when important risk trade-off decisions must be made quickly, such as in combat.

The quality of the 1-hour MEGs is limited to the quality of the source assessments. Those assessments are, in turn, limited by the quality of the database and how recently the assessments were performed. AEGLs can be assumed to be of higher quality because they were developed recently, had more data to consider, and are extensively peer-reviewed. However, they are few in number. That makes it important for TG-230 and RD-230 to be "living" documents that incorporate new values as soon as they become available. For example, RD-230 mentions that interim and proposed AEGLs were used. It is necessary to track the final versions and to ensure that the final AEGL values are incorporated into the documents. As chemicals are selected for AEGLs development, USACHPPM should give priority consideration to those chemicals likely to be found in major theaters of operations.

The 1-hour air MEGs are based on different sources. To facilitate making updates, it would be useful to document the existing guideline used and the date it was established in the supporting reference tables. In addition, the entries in air MEG tables provided in TG-230 should be reviewed to ensure that they describe specific end points of interest. Currently, some entries in the tables provide only a broad description of the critical-study end points (e.g., systemic, irritation). Sometimes the entries do not describe the end points at all (e.g., "based on a slightly higher incidence of nasal tumors in rats," "based on extrapolation of acute animal data and limited evidence in humans").

8-Hour Air MEGs

Derivation

RD-230 indicates that the exposure duration of 8 hours was selected to be consistent with brief exposures. The corresponding MEGs represent levels below which no significant adverse health effects are expected and above which the probability of adverse health effects is increased. The 8-hour MEGs incorporate the assumption that exposures will be continuous. The hierarchy used to select sources to establish the 8-hour MEGs was (1) AEGL-1 values and (2) Threshold Limit Values (TLVs), which are developed by the American Conference of Governmental Industrial Hygienists (ACGIH). The TLVs are developed to protect workers against the effects of a working lifetime of exposure (8 hours/day, 5 days/week, and 50 weeks/year for a working lifetime). RD-230 used the TLVs for 8-hour exposures when no AEGLs were available. In a number of cases, the 8-hour air MEGs are the same as the 1-hour air MEGs for minimal effects.

Evaluation

Because relatively few 8-hour AEGLs have been derived, TLVs are the preferred sources for the 8-hour MEGs. ACGIH criticizes the direct use of TLVs for purposes other than those intended. However, feasible alternatives will not exist until more AEGLs are established. TLVs are concentrations expected to be relatively safe for worker populations exposed intermittently (8-hour workdays) for a working lifetime. Thus, their direct application to a single 8-hour period is likely to be protective. Whether that approach is overly protective depends on the database and the calculations that were used to set the particular TLV. TLVs do not use standardized formulas, so it would be difficult to determine the likely margins of conservatism that were used to establish them. Determining what modifications are necessary to create an 8-hour MEG for continuous exposure from a TLV designed to be protective for intermittent exposure over a working lifetime is even more difficult. But, until revisions can be made, this approach is the most feasible, however overprotective. It appears that no consideration was given to making adjustments for the higher inhalation rate of deployed personnel; it might be appropriate to make those adjustments for 8-hour exposures. In the future, it would be useful to examine the data underlying the values selected for MEGs to determine whether the original data included exposure durations closer to 8 hours and therefore might be more appropriate than the calculations used by other agencies for other purposes.

14-Day MEGs

Derivation

The 14-day air MEGs incorporate the assumption that exposures will be continuous, recognizing the limited likelihood of that in the real world and that simplifications are essential to create workable guidelines. The hierarchy for selecting the sources of the 14-day MEGs was (1) continuous exposure guidance levels (CEGLs), (2) minimal risk levels (MRLs), (3) TLVs, and (4) special considerations. CEGLs were developed by the NRC (1986) for exposures to military personnel lasting up to 90 days. They are intended to prevent serious or permanent effects in a healthy male population and do not include consideration of susceptible subpopulations. The 14-day MEGs consider the possibility that increasing concentration or duration could increase the potential for delayed or permanent disease (e.g., kidney disease or cancer).

MRLs are developed by the Agency for Toxic Substances and Disease Registry (ATSDR) for noncancer effects. An MRL is an estimate of the daily human exposure to a hazardous substance that is likely to be without appreciable risk of adverse health effects over a specified duration of exposure. These estimates are intended to serve as screening levels to identify contaminants and potential health effects that might be of concern at hazardous waste sites. ATSDR creates MRLs for acute (1-14 days), intermediate (14-364 days), and chronic (365 days or longer) exposures.

Although RD-230 states that the TLVs were not considered protective for continuous exposures of over 24 hours to 14 days, the TLVs were extrapolated down from working lifetime values to 14-day continuous exposure values. TLVs for "systemic" or "mixed-acting" substances were adjusted by a factor of 5 days/7 days, a ventilation factor of 10 m^3/20 m^3 (10m^3 is the worker 8-hour default factor) with another calculation of 20 m^3/29.2 m^3 to account for the military person's increased ventilation rate (see Chapter 3) (equaling 10 m^3/29.2 m^3) , and an uncertainty factor of 10 to account for the uncertainty of extrapolation from intermittent to continuous exposure (see Equation 5-1).

$$\text{14-day MEG} = (\text{TLV} \times 5\ \text{days}/7\ \text{days} \times 10\ \text{m}^3/29.2\ \text{m}^3) \times 0.1 \qquad (5\text{-}1)$$

The TLVs for irritants were not adjusted because they are assumed to be mostly concentration-dependent.

USACHPPM determined that more than 24 hours of continuous exposure to chemical warfare agents (CWAs) is unlikely. Therefore, no MEGs were established for CWAs for periods greater than 1 day. Twenty-four-hour MEGs for CWAs were derived by linear extrapolation from the 8-hour MEGs.

Evaluation

The CEGLs were derived assuming 90 days of continuous exposure, so it is likely that they are conservative. Because many were published in the late 1980s, some of them could be out-of-date. Furthermore, CEGLs were developed for use by the Navy on submarines and, therefore, the target population was assumed to be exclusively male. Thus, female reproductive end points and developmental toxicity were not considered in setting the CEGLs.

The acute MRLs were calculated on the basis of exposure durations of 1-14 days. Thus, they are reasonably targeted for duration; however, they include UFs for susceptible groups. The MRL-based calculations for MEGs do not appear to include adjustments for military ventilation rates. USACHPPM should make those adjustments.

For the 14-day values based on TLVs, adjusting the TLVs for the change from 8 hour/day, 5 day/week to 24 hour/day, 7 day/week and for the higher breathing rates of military personnel (i.e., 14-day air MEGs = TLVs $\times$ 5 days/7 days $\times$ 10 m^3/29.2 m^3) is reasonable for systemic chemicals when dose rate is not the determining factor and only total dose dictates effects (Gaylor 2000). The calculations assume that a C (concentration) $\times$ t (time) = k (total exposure) relationship holds for systemic effects. The basis for the 29.2 m^3/day ventilation rate is reasonable, although it contains several assumptions.

Because TLVs are intended to be protective over a worker's lifetime, extrapolating 14-day continuous exposures from 8-hour TLVs introduces significant uncertainty. A UF of 10 was used to extrapolate from intermittent to continuous exposure. EPA and ATSDR make similar extrapolations for RfCs and MRLs, respectively, but do not use a UF. A weak justification is offered in RD-230, which says that some health effects have been observed in some workers at the TLV levels, without further specification. However, the UF of 10 is unduly conservative. Typically, concentration is more important than duration in the $C \times t$ equation (beyond acute lethality). When a guideline for intermittent exposure is converted to one for continuous exposure, it becomes more conservative. For example, consider an intermittent TLV of 0.5 mg/m^3 for 8 hour/day, 5 days/week for 260 days/year, or 2,080 hours of exposure. That is equivalent to a total exposure (k) of 1,040 mg-h/m^3. When that value is converted to a continuous exposure (24 hours/day for 365 days, or 6,360 hours), the comparable C would be 0.16 mg/m^3 (i.e., $C = k \div t$). In other words, if $C \times t = k$ operates, an intermittent inhalation of 0.5 mg/m^3 is equivalent to a continuous inhalation of 0.16 mg/m^3. Therefore, the 0.16 mg/m^3 has built-in conservatism that is appropriate. Applying an additional UF of 10 would result in a guideline of 0.016 mg/m^3, which is overly conservative.

The 14-day air MEGs are difficult to develop because most of the source materials have different exposure durations and use different assessment methodologies. Table 5-1 summarizes the sources and highlights the differences in the portions of lifetime protected and the adjustments that were made to the source material. It would be advisable to check the reference data of the sources to determine to what degree the databases were founded on studies approximating the duration of military interest. If appropriate, other data might be used to derive MEGs. Also, it would be best to use a standard approach to applying adjustments across all values. For example, adjustments for military ventilation rates should be used in all the MEGs.

1-Year Air MEGs

Derivation

The one-year air MEG is defined by USACHPPM (2002) as "The airborne concentration for a continuous exposure up to 1 year (365 days, 24

TABLE 5-1 Sources for 14-day Air MEGs

Source Reference	Exposure Duration and Frequency	Portion of Lifetime Protected	USACHPPM Adjustments in 14-Day Air MEGs
CEGL	90 days, continuous	90 days	No adjustments were made
MRL	1-14 days, daily	1-14 days	No duration or ventilation rate adjustments were made
TLVs	8 hours/day, 5 days/ week	Working lifetime	Adjustments to continuous 14-day and military ventilation rates

hours/day) that is considered protective against all health effects including chronic disease and increased risk to cancer (i.e., cancer risk greater than 1 $\times 10^{-4}$). No performance degradation or long-term health consequences are expected with exposure at or below this level. Increasing concentration and/or duration could increase the potential for delayed/permanent disease (e.g., kidney disease or cancer)." The 1-year MEGs were not designed to address continuous exposure exceeding 1 year.

Inhalation reference concentrations (RfCs) for noncarcinogenic effects, air unit risks, or inhalation cancer slope factors (CSFi) from the EPA's Integrated Risk Information System (IRIS) and the Health Effects Assessment Summary Tables (HEAST) were selected to derive preliminary long-term MEGs (PMEG-L). When those EPA sources were not available, additional sources, including TLVs and MRLs, were used with additional adjustments (see discussion below). For carcinogenic polycyclic aromatic hydrocarbons (PAHs), provisional EPA values were used; they included toxicity equivalence factors relative to benzo(a)pyrene. The air-MEG selection was based on the following hierarchy: (1) PMEG-L, (2) TLV-adjusted, and (3) MRL-adjusted. If significant (more than an order of magnitude) discrepancies between those values existed, USACHPPM reviewed the data and selected the final 1-year air MEGs.

Derivation of PMEG-Ls

USACHPPM developed military "noncancer" risk concentrations (MRCs) and cancer risk concentrations (MCRCs) using a method similar to that used in the derivation of EPA's Region III risk-based concentration values, which are consistent with risk-assessment guidance for Superfund. The cancer and noncancer values were compared, and the lower one (i.e., the more protective one) was identified as the PMEG-L.

Because a 1-year exposure duration is of interest for noncancer risks, the first-choice sources were the subchronic RfCs in HEAST; if those were not available, chronic values were used. Subchronic is defined as one-tenth of the average lifespan, or 2 weeks to 7 years, and chronic is defined as more than 7 years. To derive MRCs, RfCs (in units of milligrams per cubic meter [mg/m^3]) were converted to reference doses (RfDs, in milligrams per kilogram per day [mg/kg/day]) by multiplying the inhalation rate of 20 m^3/day and dividing by 70 kg, the average weight for adults. With the target hazard quotient (THQ) set at 1, a backward calculation was performed to derive the MRCs using the following assumptions: (1) body weight (BW) of 70 kg, (2) military inhalation rate (IRA) of 29.2 m^3/day, (3) exposure duration (ED) of 1 year, (4) exposure frequency (EF) of 365 days/year, and (5) average time (AT) of 365 days (see Equation 5-2).

$$MRC = \frac{THQ \times RfD \times BW \times AT}{EF \times ED \times IRA} \qquad (5\text{-}2)$$

EPA's CSFs were used as a basis for deriving the MCRCs. Those unit cancer risks also were converted from risk per microgram per cubic meter to risk per milligram per kilogram per day, assuming a body weight of 70 kg and an inhalation rate of 20 m^3/day. To calculate the MCRC, the target cancer risk (TCR) was set at 1×10^{-4} and the following assumptions were made: (1) BW = 70 kg, (2) IRA = 29.2 m^3/day, (3) ED = 1 year, (4) EF = 365 days/year, and (5) AT = 25,550 days (70 years × 365 days) (see Equation 5-3).

$$MCRC = \frac{TCR \times BW \times AT}{EF \times ED \times IRA \times CSF} \qquad (5\text{-}3)$$

Adjustments of TLVs and MRLs

When TLVs were used as the sources for the 1-year MEGs, they were adjusted to account for the military person's assumed respiratory rate and for uncertainties associated with extrapolating values for intermittent exposures to continuous exposures. For extrapolation, a UF of 10 was applied. However, the TLVs for irritants were not duration-adjusted.

Intermediate MRLs (15-364 days) were given preference over chronic MRLs. The MRLs were adjusted to account for the assumed military inhalation rate.

Evaluation

Quality of the Source References

EPA, ACGIH, and ATSDR sources are appropriate. IRIS is the only fully official set of EPA assessments. HEAST includes some values that have not been agreed on, some that have not been peer-reviewed, and some that have been removed from IRIS because of quality problems. Also, the use of provisional values for PAHs that are close to 10 years old suggests that some significant uncertainties in the available data have not been addressed. Furthermore, many of the current IRIS values, TLVs, and MRLs are out-of-date, and some of them are obsolete because of newer information. Although it is not feasible for DOD to revise all the source data, potential problems associated with using those sources should be recognized and stated.

EPA Region IX values were used as sources for many of the 1-year MEGs; however, those EPA values were created using a complex process and they have little apparent worth for the MEGs. The rationale offered by USACHPPM for the MRC and MCRC adjustments from milligrams per cubic meter to milligrams per kilogram per day appears to be numerically driven, and ultimately the additional conversion factors cancel themselves out. That conversion, however, was not applied to MRLs and TLVs. In developing the RfC and cancer unit risks, EPA made decisions to use milligrams per cubic meter or micrograms per cubic meter as the units for the durations of interest (typically continuous exposure for 70 years). Those units are used in the underlying research studies and are the units that would eventually be used in regulations. Although cubic meters of air breathed per day has a relationship to body weight, the convention of measuring exposure in milligrams per cubic meter is more accurate than the milligrams per kilogram per day. Conversion from an RfC or a cancer unit risk to milligrams per kilogram per day at the level of the individual studies would introduce unnecessary uncertainties.

The subcommittee recommends that USACHPPM consult additional sources of guideline values, such as the World Health Organization (WHO 2001) and the State of California. WHO (2001) has health-based guidelines for 35 air pollutants, and most were derived using expert judgment and include consideration of susceptible populations. The State of California has hundreds of values that were derived following a standard procedure similar, but not identical to that used by EPA.

The subcommittee recommends that HEAST values *not* be used to derive long-term air MEGs, because the quality of those assessments is not as strong as that of the other guidelines. For the other sources, the date of the original assessment should be provided in the tables in RD-230 to indicate the degree of potential obsolescence of the source material.

Inhalation Adjustment Factor

Most of the starting values were adjusted from EPA, ACGIH, or MRL ventilation defaults of 20 m^3/day or 10 m^3/day to the military ventilation rate of 29.2 m^3/day. The military default rate is based on a series of measurements, scenario estimations, and judgments. No default rate is or will ever be perfect, so the rate should be judged relative to its purpose of providing an appropriate level of protection for the population of concern. USACHPPM evaluated all the components of the military inhalation rate assumption, and on the basis of limited information for the types of activity likely to be performed, concluded that it is reasonable (see Chapter 3). However, the RfC methodology used to derive several of the MEGs includes a dosimetric extrapolation from animals to humans that considers ventilation rate. The implications of that are likely to be greater for reactive gases and some particles. The dosimetric model is based on a ventilation rate of 20 m^3/day, and a rate of 29.2 m^3/day would alter the pattern of respiratory tract deposition. RfCs are based on regional deposited dose when that method is supported by the data. Thus, ventilation can influence the RfC, depending on the specific circumstances. For example, a reactive gas with an assessment based on nasopharyngeal doses might be impacted. That possible impact is unlikely to have a major influence on the MEGs, but it bears consideration if there is an attempt to go back to the original data and recalculate the MEGs.

Applying a UF of 10 to the TLVs

Extrapolating the TLVs from intermittent to continuous exposures is acceptable for nonirritants because incorporating the area under the exposure curve is scientifically appropriate and is routine practice in most assessments (e.g., RfC, MRLs). However, applying a UF of 10 is not supportable, as discussed earlier (see "14-Day Air MEGs"). The MEGs should therefore be revised.

Varying Exposure Durations

One overarching problem is that the 1-year air MEGs are based on source references with varying exposure durations. Table 5-2 summarizes the exposure durations and frequencies associated with the sources used by USACHPPM in developing the 1-year air MEGs.

USACHPPM mentioned that they used subchronic RfCs when they are available. However, the chronic RfCs and unit cancer risks, which are both based on lifetime exposure, were also used. For carcinogenic agents, the MCRCs are derived by averaging the 70-year cumulative lifetime dose limit

TABLE 5-2 Exposure Durations in Source References

Source References		Duration	Frequency
PMEG	Noncancer (RfCs)	1/7 of lifetime or lifetime	Daily
	Cancer (unit cancer risks)	Lifetime	Daily
MRL	Noncancer	15-364 days	Daily
TLV	Noncancer and cancer	Working lifetime	8-hour day, 40-hour workweek

(given a unit cancer risk and lifetime risk of 1×10^{-4}) over a 1-year exposure duration. That derivation is appropriate and is protective for exposure durations up to 1 year.

One problem with using the chronic RfCs as starting points for 1-year values is that in some cases those RfCs are based on subchronic effects (e.g., a 90-day exposure study of laboratory rats), and a UF of 10 was applied in the original derivation to extrapolate from subchronic to chronic exposure. Because subchronic exposures are of direct interest to MEGs, there is no need for that particular UF. For example, the IRIS RfC for acetaldehyde has a composite UF of 1,000, including a factor of 10 for subchronic to chronic extrapolation. Thus, that starting point would be over-protective for a 1-year exposure scenario.

In other cases, the original assessment, following the established methodology, would have used lifetime exposure studies, if available and of quality, to derive an RfC, and would only have used subchronic studies to enhance understanding of the chemical. If the goal was a 1-year exposure guideline, high-quality subchronic studies likely would be used for a derivation. Some RfCs might rely on robust chronic studies when subchronic data are inadequate. Those RfCs would be used appropriately as input for a 1-year MEG. An added complication is that the assessment methods of other agencies are changing. For example, benchmark dose methods are becoming more common. Thus, approaches that were adequate for the underlying methods might need to be revisited when USACHPPM revises the MEGs.

For 1-year air MEGs that are based on adjusted TLVs, there is an additional interpretative issue, because the TLVs address worker-lifetime exposure. Therefore, the adjusted TLVs would be protective for exposure durations much longer than 1 year. Although that added level of protectiveness might be appropriate in light of the potential for repeated and multiple deployments, the lack of consistency in adjusting source reference exposure durations to fit the MEGs poses interpretation challenges. Efforts should be made to increase consistency in future revisions of MEGs.

Specific Chemical MEGs

A few 8-hour, 14-day, and 1-year air MEGs were selected outside of the established hierarchy. The MEGs for benzene and toluene were based on the TLV-adjusted rather than the MRLs and PMEGs. The ethyl benzene MEG was based on the MRL-adjusted rather than the PMEG. All of those decisions were made to avoid relying on conservative UFs. Related issues arose for styrene, *n*-hexane, and xylene. Several PAHs were considered, but inhalation toxicity data were lacking for seven of them. Oral RfD data were extrapolated, and quantitative structure-activity relationship methods were used to derive inhalation values for those seven PAHs. The oral-to-inhalation extrapolation for PAHs is fraught with uncertainty, but there are no reasonable alternatives.

The MEG for naphthalene was based on the MRL-adjusted because that source addressed subpopulations with glucose-6-phosphate dehydrogenase (G-6-PD) deficiencies that did not appear to be addressed by the TLV. That subpopulation could be more vulnerable to oxidant exposures. There are two issues: (1) do people with G-6-PD deficiency truly need more protection (they would in theory, but how good is the evidence for likely impact), and (2) if they do, protection against more than naphthalene (e.g., other oxidants) might be indicated. USACHPPM needs to analyze that further.

Ambient Air Quality Criteria Pollutants

EPA health-based national ambient air quality standards (NAAQS) exist for six pollutants: ozone (O_3), nitrogen dioxide (NO_2), sulfur dioxide (SO_2), lead, particulate matter (PM_{10} and $PM_{2.5}$), and carbon monoxide (CO). These pollutants are important because they are ubiquitous and capable of causing adverse health effects in susceptible individuals at high ambient levels. The NAAQS have a variety of durations (some are for 24 hours, some are annual) and are set to protect susceptible subpopulations. Each of the standards also has a descriptive category of air quality (the pollutant standard index [PSI]) developed by EPA to provide precautionary summaries about health effects to nontechnical audiences.

One-year MEGs were developed for the six criteria air pollutants. Annual mean, quarterly averages, and 24-hour NAAQS were considered when available. Linear extrapolation was used for substances that only had 8-hour average standards. TLVs also exist for these criteria pollutants, so the TLV-adjusted was compared with the NAAQS. The rationale for choosing one versus the other was not described.

NAAQS are based on extensive databases and are routinely updated by EPA. It would be relatively easy to re-evaluate these databases with an eye towards identifying subpopulations that might be represented in the deployed population, thereby getting a more accurate indication of potential risks from these ubiquitous pollutants. EPA has several experts on criteria pollutants who could be approached for assistance.

The rationales for choosing the long-term MEGs for the criteria pollutants need further discussion. For example, why was the NAAQS for NO_2 (0.1 mg/m^3) chosen over the TLV-adjusted (0.14 mg/m^3) when the NAAQS is set to protect children and the TLV is set to protect adults? For O_3, the MEG is 0.052 mg/m^3, the 8-hour NAAQS is 0.157 mg/m^3, and the TLV-adjusted is 0.004 mg/m^3. Background levels for O_3 in many areas of the world are above the long-term MEG. Thus, the MEG would likely be exceeded frequently in those areas that have a lot of sunlight and have precursors (NO_x and VOCs). This issue is complicated by the fact that low levels of O_3 (as low as 0.12 mg/m^3 under heavy exercise conditions for about a half hour [EPA 1996a]) can affect the pulmonary function and exercise performance of young healthy athletic adults. It would be advisable to re-evaluate the MEGs for O_3 to ensure that they appropriately protect the forces. The SO_2 annual NAAQS is 0.365 mg/m^3, which includes consideration of susceptible subpopulations, but the 1-year MEG is 0.13 mg/m^3 and is based on an adjusted TLV. The unadjusted TLV for SO_2 is 5.24 mg/m^3. From a mathematical and procedural viewpoint, the rationale for that choice is clear. However, selecting a military value *lower* than the NAAQS should be re-evaluated, especially considering the more robust health of the military population. As mentioned above, the most straightforward approach to dealing with the criteria air pollutants might be to consult with EPA and address the military population explicitly.

The subcommittee recommends that the MEGs for criteria pollutants be reconsidered relative to their applicability to deployed forces. This will require evaluating the chemical database for data relevant to deployed forces and deriving MEGs from those data. The following comments show why certain NAAQS are inappropriate bases for the MEGs (Graham et al. 1999). Some specific comments are offered in Appendix A.

- As mentioned above, the NAAQS are set to protect susceptible subpopulations. In the case of O_3, the susceptible populations include healthy young adults exercising outdoors, which makes the NAAQS relevant to the troops. In contrast, other NAAQS are not directly relevant to deployed forces, even assuming some degree of health impairments among the military population (e.g., mild asthma).
- The NAAQS for CO are set to protect angina patients (or others with forms of coronary artery disease) who exercise; therefore, they are not applicable to deployed troops. The goal of the current CO standards is to maintain blood carboxyhemoglobin (COHb), a biomarker of effects, below 2.1%. Average COHb levels in smokers who smoke 1-2 packs per day are about 4%. CO can have effects on healthy individuals and fetuses, but at higher levels than the NAAQS. Those effects include reduced maximal exercise duration and central nervous system effects (e.g., decrements in hand-eye coordination) that could be of concern during deployments.
- The NO_2 annual-average NAAQS is based on effects on 5- to 12-year-old children. EPA does not have a short-term NAAQS for NO_2 because if the annual NAAQS is met, there is almost no likelihood of short-term concentrations rising to levels of concern for other healthy or susceptible populations. However, that analysis is based on ambient air quality patterns in the United States. The potential exists that other countries with different air quality patterns might have short-term excursions that would affect asthmatic individuals. WHO (2002) set a short-term air quality guideline that might be valuable in this instance.
- The NAAQS for SO_2 (24-hour and annual average) are set to protect children and people with pre-existing lung disease (especially asthma). Short-term (e.g., 10-min) exposures in exercising asthmatic individuals could cause bronchoconstriction. However, a significant prevalence of asthma among deployed troops is unlikely.
- The NAAQS for PM are set to avoid increased mortality, predominantly from cardiovascular and respiratory causes, although children are also at risk for respiratory morbidity. Healthy young adults are not likely affected, even at high ambient levels.

DRINKING WATER GUIDELINES

This section reviews USACHPPM's considerations for contaminated drinking water. Packaged and treated water is not always available during deployments. Therefore, guidelines are necessary to evaluate the quality of local water sources.

How Water MEGs Were Derived

USACHPPM reviewed drinking water standards from EPA and other organizations to identify starting points for calculating 5-day, 2-week, and 1-year water MEGs. Table 5-3 provides the health-effects definitions used for determining the water MEGs. As with the air MEGs, certain adjustments were performed to make the values more relevant to the military population. For example, the daily water consumption rates of deployed personnel are much higher (between 5 and 15 liters [L]) than those of the general population (which are assumed to be about 2 L).

The hierarchy used to choose references on which to base the water MEGs was set as follows:

1. DOD tri-service military field drinking water standards. Field drinking water standards (FDWS) were developed for military personnel primarily to prevent performance degradation on the battlefield. The military water consumption rate was assumed, and no uncertainty factors were used to protect susceptible subpopulations. FDWS were used to derive water MEGs for arsenic, chloride, cyanide, lindane, magnesium, and sulfate as well as the CWAs sulfur mustard, lewisite, nerve agents (GA, GB, GD, and VX), BZ, and T-2 toxins. For CWAs, only 24-hour MEGs were derived, because the likelihood of CWA exposures extending beyond a 24-hour period is small.
2. EPA health advisories (HAs). HAs were used to derive about 50% of the 1-year water MEGs. HAs are guidelines provided for exposure durations of 1 day, 10 days, long-term, or lifetime. Long-term is defined as less than 7 years, and those values are based on a 70-kg adult consuming 2 L of water each day. Adjustments were made to compensate for the higher drinking water consumption rate of military personnel.
3. MRLs. MRLs were used in the absence of drinking water guidelines (like MCLs or HAs). Intermediate levels were chosen (15-364 day potential exposure). Because MRLs are expressed as daily human doses in milligrams per kilogram per day, it was necessary to convert the values to corresponding water concentrations by assuming a 5 L/day consumption rate and an adult body weight of 70 kg.
4. Reference doses (RfDs). Subchronic or chronic RfDs were used as the basis of water MEGs when no other existing long-term health guidelines were available. About 20% of the long-term water MEGs were calculated using the RfDs. The RfDs were obtained from EPA's HEAST, IRIS, or Region III risk-based concentration (RBC) tables. Because RfDs are expressed as daily human doses in units of milligrams per kilogram per day,

TABLE 5-3 Definitions of Health Effects Associated with Water MEGs

Exposure Duration	Health Effect	Health Effects and Performance Degradation[a]
5 days, 5 or 15 L/day	Minimal to Nonsignificant	The drinking water concentration for a continuous daily consumption of either 5 L/day or 15 L/day for up to 5 days that should not impair performance and is considered protective against significant noncancer effects. Increasing concentration and/or duration could result in performance degradation, need for medical intervention, or increase the potential for delayed or permanent disease (e.g., kidney disease or cancer)
14 days, 5 or 15 L/day	Minimal to Nonsignificant	The drinking water concentration for a continuous daily consumption of either 5 L/day or 15 L/day for up to 14 days that should not impair performance and is considered protective against significant noncancer effects. Increasing concentration and/or duration could result in performance degradation, need for medical intervention, or increase the potential for delayed or permanent disease (e.g., kidney disease or cancer)
1 year, 5 or 15 L/day	Nonsignificant to None	The drinking water concentration for a continuous daily consumption of either 5 L/day or 15 L/day for up to 1 year that should not impair performance and is considered protective against health effects including chronic disease and increased risk to cancer (i.e., cancer risk greater than 1×10^{-4}). Increasing concentration and/or duration could increase the potential for delayed or permanent disease (e.g., kidney disease or cancer)

[a]Sensitive individuals could be predisposed to toxic effects and, therefore, could be more susceptible. If available scientific evidence regarding susceptible subpopulations exists for a particular chemical, then that information is provided in the guideline tables.
Source: USACHPPM 2000a.

it was necessary to convert the values to corresponding water concentrations by assuming a 5 L/day consumption rate and an adult body weight of 70 kg.

To assess whether the long-term MEGs were protective against cancer, USACHPPM compared risk-specific (1×10^{-4}) concentrations of carcinogenic chemicals with the corresponding long-term water MEGs. The risk-specific concentrations were obtained from EPA's drinking water regulations and HAs or from IRIS, and they all assumed a lifetime of exposure. Those values were adjusted to estimate concentrations in water that would pose the same cancer risk for an exposure of 1 year and to account for the military water consumption rate. If the adjusted cancer-risk concentration was equal to or greater than the corresponding long-term MEG, the MEG was considered to be protective against cancer. When the adjusted risk-specific value was less than the long-term MEG, the MEG was replaced with that value.

Evaluation of and Recommendations on the Derivation of Water MEGs

Selection of Chemicals and Existing Standards

The chemicals for which water MEGs were developed were selected on the basis of USACHPPM's review of existing water contaminants, including chemicals listed in Technical Bulletin, Medical 577 (U.S. Department of the Army 1999); chemicals detected during water sampling in Bosnia; and compounds listed as high priority in RD-230. Although that is a good start, USACHPPM should consider periodically updating the list of chemicals with other compounds that are likely to be present, such as industrial mixtures, like gasoline and diesel fuel, and newly identified contaminants.

USACHPPM used relevant existing guidelines set by other agencies as starting points for deriving water MEGs. The major difficulty with that approach is that the target populations for the existing guidelines were usually different from the deployed population, which means that some of the assumptions and considerations used might not be relevant to military guidelines. This is discussed in detail in Chapters 2 and 3 and above in the discussion of air MEGs. Other problems include ensuring that the values are properly recorded and updated and that they gain acceptance within the scientific community. For example, the water MEG for lead is based on a WHO guideline that was incorrectly recorded by USACHPPM as 0.05

mg/L instead of 0.01 mg/L. In addition, several exposure values taken from EPA's HEAST database were not peer-reviewed and are not as accepted as other guideline values.

Time Frames

Water MEGs are set for short-term exposures of 5 days and 2 weeks and for a longer period of 1 year. Long-term MEGs were not set for CWAs because their chemical and physical characteristics make long-term exposures implausible. These time frames seem appropriate for the military population, whose exposures might range from a limited number of days up to a year.

Route of Exposure

The guidelines reflect drinking water exposures only. No bathing, dishwashing, or other potential nonpotable water exposures were considered. It appears that USACHPPM considered exposure to water contaminants by inhalation or dermal exposure to be much less than would occur from ingestion of water. In the case of dermal exposures, the subcommittee performed rough risk calculations using data for some volatile and semivolatile organic compounds, such as trichloroethylene and benzo(a)pyrene, to evaluate dermal exposure during showering and agrees with USACHPPM that risk from water consumption generally subsumes that from dermal absorption. This was true even for chemicals known to have high potential for skin penetration (evaluated by comparing penetration constants [Kp's]). If the Kp was greater than or near unity (as was the case for certain PAHs), those moieties posed dermal risk close to but not greater than drinking water risk. However, it was unclear whether inhalation of volatile chemicals during showering would be as minimal. This is an issue that needs further consideration by USACHPPM. It might also be worthwhile for USACHPPM to consider more unusual exposures to water contaminants, such as those that might occur in water emersion scenarios.

Water Consumption Rates

Water MEGs are based on specific exposure conditions defined by estimated daily water consumption rates. Five liters per day is often used as the default military water consumption rate, but in dry, arid climates that rate could be as high as 15 L/day. These high rates have been validated and established in Army doctrine (U.S. Department of the Army 1999) and are consistent with reports from the Israeli Defense Forces and U.S. Army Medical Services officers in the Mojave Desert (Henry 1985).

SOIL EXPOSURE GUIDELINES

This section reviews the technical guidance in RD-230 for deriving guidelines for chemical hazards in soil. In contrast with the MEGs for air and water, soil MEGs are derived for only long-term exposures.

How Soil MEGs Were Derived

Risk-based soil screening levels are derived by using selected target risk levels, assumptions about exposure routes and characteristics, and toxicity values to calculate acceptable exposure concentrations. For soil contaminants, three exposure routes are typically considered: inhalation of resuspended soil particulates; ingestion of soil; and dermal absorption of chemicals from soil adhered to skin. When all three exposure routes are included, the soil MEG represents a soil concentration that will not cause unacceptable health risks even when the three exposure routes are combined. It should be noted that soil MEGs are the only MEGs that consider multiple routes of exposure. For VOCs, the soil saturation concentrations were used as the soil MEGs when they exceeded the health-based values.

Soil MEGs were derived in a manner consistent with the derivation of risk-based screening levels for the general population with adjustments added to better represent the characteristics of deployment exposures. The chemicals selected for soil MEG development were described by USACHPPM as "consistent with those used to develop drinking water guidelines." The rationale for the selection process is that ingestion is the primary exposure route for both soil contaminants and drinking water contaminants.

The target risk levels for soil MEGs are consistent with the target risk levels for air and water long-term MEGs (a target cancer risk of 1×10^{-4}), although a hazard quotient of 1 was used for noncancer end points. For chemical mixtures, carcinogenic effects are assumed to be additive, whereas for noncancer effects, target organs should be evaluated to determine if additivity is likely.

Three EPA methods for deriving risk-based soil screening levels were considered for use in developing soil MEGs. The EPA methods included the Office of Solid Waste and Emergency Response soil screening levels (SSLs) (EPA 1996b), the EPA Region III RBCs (EPA 1999a), and the EPA Region IX preliminary remediation goals (PRGs) (EPA 1998). These three methods have some differences in assumed exposure scenarios, routes, and exposure characteristics. All three methods use similar approaches to develop soil standards for residential land use, but only the RBCs and PRGs included industrial soil goals at the time the MEGs were developed. Exposures of deployed personnel will be more similar to industrial exposures than to residential exposures. The EPA Region IX PRG method was selected for modification by DOD because "it results in the most conservative soil concentrations since it includes more exposure pathways than either the SSL or the RBC methodology." Adjustments were made to the selected model to more accurately represent deployment conditions.

In choosing inhalation toxicity values, USACHPPM used the same hierarchy as was used to derive the air MEGs. For oral toxicity values, CSFs from IRIS and HEAST were used for carcinogens, and the water MEG values were used to back calculate oral RfDs. Dermal toxicity values were derived by adjusting the oral toxicity values to represent absorbed dose instead of ingested dose whenever gastrointestinal absorption data were available.

Critical assumptions used to estimate oral, inhalation, and dermal exposures include soil ingestion rate, inhalation rate, skin surface area, skin adherence factors, and dermal absorption. In RD-230, an average soil ingestion value of 265 mg/day was derived by assuming that soldiers have equal numbers of high ingestion and low ingestion days while deployed. A high ingestion value of 480 mg/day was assumed to be associated with activities such as digging or crawling on the ground, and 50 mg/day was assumed to be the mean ingestion rate on days when troops do not engage in such activities. The daily inhalation rate of 29.2 m^3/day that was used for derivation of air MEGs was also used for the soil MEGs. Dermal exposure assumptions were based on a 90th percentile value for adult male skin surface area, a soil adherence factor of 1.0 milligram per square centimeter (mg/cm^2), and dermal absorption values of 1% for inorganic chemicals and 10% for organic chemicals.

The soil MEG for lead was treated as a special case. RD-230 provides a soil MEG for lead derived on the basis of an EPA modification of the Bowers adult lead model (Bowers et al. 1994; EPA 1996c). EPA greatly increased the conservatism of the model by increasing both the biokinetic slope factor (from 0.375 to 0.4 micrograms [µg] per decaliter [dL] per µg/day) and the absorption factor for lead in water and diet (from 0.08 to 0.20)

(Bowers and Cohen 1998). DOD has modified the model further by increasing the target blood lead concentration from 10 μg/dL for the fetus to 30 μg/dL for an adult worker. The soil ingestion was also increased from 50 mg/ day to 265 mg/day.

Evaluation of and Recommendations on the Derivation of Soil MEGs

The general approach used to evaluate soil hazards conforms to current risk-assessment practices. However, the accuracy of the resulting MEGs depends on whether and how the guidelines will be kept up to date as EPA guidance and the underlying science continues to evolve. There are already a number of instances in which specific assumptions used to derive the current MEGs have been (or will soon be) superceded by revised and updated guidance from EPA. The subcommittee's specific comments and recommendations are described below. Minor errors are noted in Appendix A.

Selection of Chemicals

As described above, the chemicals for which MEGs were developed were selected with the assumption that ingestion is the primary exposure route for soil contaminants. That rationale should only apply to nonvolatile chemicals, because for volatile chemicals, the primary route of exposure from soil is often inhalation of released vapors. Inhalation exposures are of particular concern for troops in trenches or in tents or buildings. Common volatile chlorinated solvents, such as tetrachloroethylene, trichloroethylene, and carbon tetrachloride, were inappropriately excluded from the soil MEGs. Elemental mercury is another common, highly volatile industrial chemical that should have a soil MEG based on potential inhalation exposures.

RD-230 states that soil MEGs might be derived for manganese and selenium in the future. Oral RfDs are available on IRIS for both of those chemicals at this time; however, deriving soil MEGs for manganese and selenium should not be a high priority, because their toxicities are generally low.

Method Selection

DOD should update its references for the three methods considered for deriving MEGs. Both the EPA Region III and EPA Region IX databases were updated in October 2002 (EPA 2002a,b), and the newer references should be cited. In addition, EPA has developed supplemental SSL guidance (EPA 2002c) that includes an industrial scenario. That guidance should be discussed in the next update of RD-230 and should be considered as a possibly more appropriate method for soil-MEG derivation. Any revised assumptions in the updated guidance should be considered for inclusion in the soil MEGs. For example, in the derivation of their particulate emission factor (PEF), EPA (2002c) has changed the value of its site-specific dispersion factor (Q/C) from 90.8 (g/m^2-s per kg/m^3) to 93.77 (g/m^2-s per kg/m^3).

As described above, the EPA Region IX PRG method was selected because it considers more exposure pathways than the other methods. However, DOD has excluded the inhalation route of exposure to volatile chemicals released from soil from the calculations for soil MEGs. Because that exposure route is probably responsible for the relative conservatism of the PRGs, excluding it is inconsistent with the rationale for selecting the PRG method.

The reason that vapor inhalation was excluded form the soil MEGs is that field sampling of air concentrations should also capture vapors released from soil. That assumption needs to be examined further. Volatile organic compound (VOC) releases from soil are dependent on temperature and other weather-related factors. It is possible that field sampling would not occur under the conditions most likely to facilitate the release of VOCs. SSLs based on vapor inhalation are also generally lower than SSLs based on soil ingestion. Evaluating VOCs in soils is further complicated by the technical difficulty of obtaining accurate measurements of VOC concentrations in soil due to vapor loss from samples. One possible approach might be to follow the SSL method of listing two separate standards—one for combined ingestion and dermal exposures, and another for inhalation exposures. Identifying exposure-route-specific soil MEGs has the added advantage of providing information that might be useful in identifying appropriate protective actions in the event of exposure. Actions appropriate for avoiding exposures to vapors released from soils will be very different from those appropriate for avoiding soil contact leading to oral and dermal exposures.

Toxicity Data

As noted above in the recommendations for the derivation of air MEGs, the subcommittee recommends that USACHPPM avoid using toxicity values listed in the HEAST tables for the derivation of MEGs. Also, USACHPPM should note the date of the underlying toxicity assessment used to derive the MEGs in the RD-230 tables.

The subcommittee noted several flaws in the adjustments made to the inhalation toxicity values used to derive the air MEGs (see comments on air MEGs, above). The inhalation toxicity values used to derive the soil MEGs should be re-evaluated in light of those comments. Once those flaws are corrected, it is likely that the inhalation pathway will not have a significant influence on the soil MEGs for semivolatile and inorganic chemicals. The adjustments made to oral toxicity values for deriving the water MEGS were judged to be appropriate, and their use in deriving soil MEGs is also appropriate.

The description of the derivation of dermal toxicity values in RD-230 is misleading and includes errors that need to be corrected. The dermal toxicity section of RD-230 cites a 1989 EPA document as a source of guidance in the development of dermal toxicity values. The supplemental guidance for dermal risk assessment is more current and is available as a public review draft (EPA 2001a); it was scheduled to be finalized in 2003. EPA (2001a) indicates that VOCs can be excluded from evaluations of dermal exposures to soil contaminants because they are released to air before being absorbed. Also, only two inorganic chemicals are currently included in the EPA guidance (arsenic and cadmium).

RD-230 should be revised to explain that the evaluation of dermal exposures requires the conversion of oral toxicity values based on intake into toxicity values based on absorbed dose. The text says "If a chemical-specific GI ABS [gastrointestinal skin absorption factor] is not available, then a default value of 100 percent is recommended." RD-230's Table E-4 shows oral absorption values of 100% for all of the chemicals listed. EPA (2001a) should be consulted as a source of oral absorption values. Generally, oral absorption of organics is greater than 50%, and EPA (2001a) has determined that no adjustment of oral toxicity values is necessary when oral absorption is 50% or greater. Many inorganics are much less completely absorbed, so the use of an assumed 100% oral absorption would not yield adequately protective values for those chemicals. For example, oral absorption of cadmium is 5% or less. The proper adjustment for evaluating risk associated with an absorbed dermal dose of cadmium would be to multiply the oral RfD of 1×10^{-3} mg/kg-day (based on ingested dose) by a factor of 0.05 to yield an absorbed dose RfD of 5×10^{-5} mg/kg-day. Using the properly adjusted RfD to evaluate dermal exposure to cadmium yields a risk estimate 20-fold greater than that yielded using the unadjusted oral RfD. Consequently, use of unadjusted oral toxicity values to assess dermal exposures could lead to substantially underestimated risk estimates. This section of RD-230 should be revised to reflect EPA's guidance.

The dermal toxicity discussion in RD-230 also notes that the absence of an ACGIH skin notation was used to preclude some chemicals from evaluation for the dermal exposure pathway. That assumption should be examined more carefully. If the ACGIH notations are based on exposures to pure chemicals, they might not be relevant for assessing the potential for dermal absorption from soil mixtures. It would be more consistent with current risk-assessment practice to select chemicals on the basis of EPA's latest guidance.

Exposure Factors

Soil contamination is assumed not to pose an immediate or severe hazard unless there are obvious, avoidable signs such as odor, discolored vegetation, or free chemical product. That assumption is appropriate for oral or dermal exposures to chemicals in soil but might not always be applicable to exposures to vapors released from soil. Vapor inhalation is not likely to be an immediate concern for outdoor, ground-surface activities, but some activities in trenches or enclosed areas such as buildings or tents might allow vapors released from soil to build up to concentrations of concern that would not be detected by ambient air monitoring. The potential for that should be addressed in RD-230, and some screening calculations should be performed to determine the potential health threats from volatile chemicals in subsurface soil.

As described above, the critical assumptions used to estimate inhalation, oral, and dermal exposures include inhalation rate, soil ingestion rate, skin surface area, skin adherence factors, and dermal absorption. The inhalation assumptions were addressed in the review of air MEGs. For soil ingestion estimates, RD-230 properly notes that there is extreme uncertainty in those values for adults because of an almost complete absence of reliable empirical data.

The discussion of dermal exposures in RD-230 needs to be updated with references to EPA's new dermal risk-assessment guidance (EPA 2001a). DOD used the 90th percentile of adult male body surface area for head, hands, and forearms in calculating dermal exposure. That assumption is in direct conflict with current EPA risk-assessment guidance, which notes that the 50th percentile population values should be used whenever 50th percentile body weight values are used, because body surface area and body weight are correlated. The estimated surface area used by DOD is 4,090

cm^2, whereas EPA used a value of 3,300 cm^2. The RD-230 value should be changed to be consistent with the EPA guidance.

An adherence factor of 1.0 mg/cm^2 was selected to derive soil MEGs. That value is much higher than the EPA value of 0.2 mg/cm^2 for industrial exposures (EPA 2001a). A number of studies have increased our database for soil adherence to skin (Kissel et al. 1996, 1998; Holmes et al. 1999), and they were used by EPA in deriving the new adherence factors. Those studies should be considered in re-evaluating adherence factors for soil MEGs. In particular, DOD should consider using high and low soil adherence values with the same frequency that high and low soil ingestion values are used. The adherence factor of 1.0 mg/cm^2 should only be applied to the activities that are assumed to have a higher soil ingestion rate (activities such as digging or crawling on the ground). The lower adherence factor of 0.2 mg/cm^2 should be applied to other activities. Because DOD assumes that the high-soil-contact activities occur 50% of the time, the average soil adherence would be 0.6 mg/cm^2.

The discussion of chemical-specific dermal absorption factors should be revised to reflect the most current guidance (see EPA 2001a). As noted earlier, VOCs can be excluded from the dermal pathway, and only a subset of inorganics is included by EPA at this time. DOD also presents specific dermal absorption factors for CWAs that were derived by assuming that the chemicals in soil first dissolve in water and then are absorbed at rates determined by their flux from water (USACHPPM 1999). The partitioning of chemical from soil to water is predicted on the basis of the octanol water coefficient (K_{ow}) and the fraction of organic carbon in the soil. The flux is determined by a model based on water solubility, K_{ow}, and molecular weight. Estimates of the hourly absorption fractions for CWAs in soil are then derived from the flux (e.g., 0.35% for GB). Those hourly fractions can add up to relatively high absorption estimates over time. For example, for a 12-hour exposure, the absorbed fraction would increase to 3.3% of the amount of GB in adhered soil.

The estimated absorption fraction for GB derived using the DOD model is at least an order of magnitude greater than values observed for chemicals that have much higher octanol water coefficients. For example, Wester et al. (1992) did a study with chlordane in soil. Chlordane was added to soil to yield a concentration of 67 parts per million (ppm) and was placed on the skin at a density of 40 mg of soil per square centimeter with 2.7 μg of chlordane per square centimeter. At 24 hours, 0.34% of the original dose was in the skin and 0.04% had penetrated the skin. Counting both the amount that was in the skin and the amount that had penetrated the skin, the calculated flux was 0.01 μg/cm^2/day (2.7 × 0.0038), which is equivalent to 0.00042 μg/cm^2/hour or 0.015% per hour (more than an order of magnitude below the 0.35% estimated for GB using the octanol water partitioning model). Adding the amount of chemical in the skin to the flux is a conservative assumption, because the chemical might evaporate or might be bound in the skin and sloughed with normal skin turnover. If the DOD model had been applied to chlordane, a flux much greater than the estimated flux for GB would have been predicted. Consequently, the discrepancy between the empirical data for chlordane and DOD's predicted values casts doubt on the validity of the DOD model.

It appears that there is not any scientific support for the assumption that partitioning to water will be predictive for dermal absorption from soil. EPA does not estimate dermal absorption of other soil contaminants using a flux, but instead uses estimates of the absorbed fraction of applied dose each day. This is a controversial subject at present, and which model is most appropriate might depend on how contact with soil occurs. EPA's model essentially assumes one soil application per day that stays on the skin all day, whereas other scientists argue that there would be continued soil contact and turnover on skin throughout activities. If soil stays on the skin for a prolonged period, the amount of contaminant absorbed per hour is likely to decrease over time, so assuming a constant flux might not be appropriate. It seems likely that there would be some turnover during activities that bring personnel in contact with soil, but it also is likely that a residual amount would stay on the skin for a prolonged period (especially during deployments).

With a few exceptions, it is implausible that CWAs would be present in soil for prolonged periods of time. Newer CWA residues would be expected to behave differently than older residues, and absorption of new residues is likely to be substantially different from absorption of old residues. Short-term soil MEGs should be considered for the volatile CWAs, and long-term MEGs should be derived only for the toxic breakdown products and for HD (which can be preserved in a coating of polymeric hydrolysis products).

For short-term CWA soil MEGs, the Army will need to break down the screening levels by exposure route to allow for field personnel to decide on appropriate protective actions. It is vital to know if the primary risk is from inhalation, dermal exposures, or soil ingestion. Soil ingestion is relatively easy to avoid, whereas inhalation can not be avoided without a respirator.

Lead

As described above, EPA increased the conservatism of the Bowers adult lead model by increasing both the biokinetic slope factor and the absorption factor (Bowers and Cohen 1998). DOD then increased the target blood lead concentration from 10 µg/dL for the fetus to 30 µg/dL for adult workers, an action that reduced the protectiveness of the model. DOD also increased the soil ingestion from 50 mg/day to 265 mg/day. It is not clear whether these and other assumptions made by DOD are justified. The 30 µg/dL target blood lead might not be protective enough for the embryo and fetus. However, many of the other assumptions are overly conservative and could result in substantially overpredicted blood lead levels. Lower lead absorption rates are supported by a validated pharmacokinetic model (O'Flaherty 1993), and a lower biokinetic slope factor was used to accurately predict blood lead levels during development of the original model (Bowers et al. 1994).

The background blood lead level used by DOD is based on a 1988-1991 survey of blood lead levels in the U.S. population. Contrary to the erroneous definition of PbB_1 as background blood lead concentrations in adult males, the values in table RD 3-7 are for women 17-45 years old, which is the appropriate receptor demographic to evaluate. More recent data suggest declines in the blood lead levels of women of child-bearing age since 1988-1991, the range of blood lead levels having fallen from 1.7-2.2 µg/dL to 1.4-1.9 µg/dL (EPA 2002d). Table RD 3-7 is missing a value for geometric standard deviation (GSD), a parameter needed to calculate soil screening levels. EPA (2002d) indicates that as background lead levels have fallen, the GSD has risen. However, the rise in the GSD appears to be an artifact of increasing proportions of nondetected values in the database (up to 25%). Therefore, EPA's (2002d) conclusions should not be accepted without a detailed critical evaluation.

The soil ingestion rate selected by DOD is very high and is highly uncertain. The adult lead model was designed to use a central tendency estimate of soil ingestion. To the extent that the DOD value is an upper-end estimate, its use will lead to substantial overprediction of blood lead levels. The model requires central tendency estimates for all parameters and applies a GSD to derive a population distribution. Use of input parameters that are not central tendency estimates is indefensible given the model's structure.

The RD-230 analysis yields a soil MEG for lead of 2,200 ppm. On the basis of observations made in communities that have elevated soil lead levels, it was concluded that 2,200 ppm is not likely to result in unacceptable exposures. The overly conservative exposure assumptions might have balanced out the unprotective target blood lead concentration to yield a reasonably protective soil MEG. Given the availability of reliable tests for blood lead levels, it might be feasible to periodically monitor blood lead levels in troops exposed to soils whose lead levels exceed the soil MEG to ensure that target blood lead levels are not also exceeded.

Consideration of Acute Toxicity

The soil MEGs are set for 1-year exposures. USACHPPM performed an analysis to verify that soil MEGs do not pose an acute risk at higher exposure rates by comparing the soil MEGs to EPA's 1-day drinking water HAs. RD-230 cites an EPA document from 1996 (1996d) for the HA values, but a 2002 update of the HAs posted on EPA's website includes some short-term HAs dated 1998 (EPA 2002e). The more recent reference should be used. Also, many of the HAs are based on assessments that are more than 10 years old. Acute toxicity values (MRLs) for many of the chemicals of interest are also available from ATSDR (2003). If those are generally more up-to-date than the HAs, it might be advisable to use them instead.

In evaluating potential acute exposures from soil, DOD focused solely on soil ingestion. As described above, the subcommittee does not agree with DOD's rationale for excluding many volatile chemicals from soil-MEG derivation. When more accurate long-term soil MEGs that consider inhalation exposures are derived for volatiles, it will also be necessary to develop short-term MEGs to evaluate potential acute risks from those chemicals.

APPLICATION OF MEGs

Deployment conditions are undoubtedly complex. Potential toxic chemical exposure scenarios are expected to be highly variable from one deployment to the next. Personnel with sufficient knowledge and field experience are needed to assess the health risks associated with these complex potential exposures by using the information provided through the MEGs. However, more explicit guidance on how to apply the MEGs is necessary to ensure that assessments and the resulting management actions are performed consistently. Particularly, guidance is needed on how to develop appropriate risk-management plans for when measured or predicted exposure concentrations exceed the MEGs and how to adequately characterize the risks in the event of exposures to the same toxic agent through multiple routes and pathways or simultaneous exposures to several chemicals.

Interpreting MEG Exceedances

As noted in Chapter 2, two separate sets of guidelines are necessary to appropriately assess chemical threats to the mission and chemical threats to force health. Under this new scheme, it would not be appropriate for the health-based MEGs to be used in conjunction with the military's mission risk assessment matrix to evaluate mission risks, as is currently recommended in TG-230. Rather, the subcommittee envisions that MEGs will be used as guidelines to assess health risks and the potential risk-management options for reducing or eliminating those risks. That information would then be considered by decision makers in conjunction with mission-related risk assessments. For example, in cases where some level of health risk is accepted to complete the military objective, MEGs could be used to determine the medical follow-up responsibilities of DOD. The subcommittee recommends that DOD develop a risk-management framework that focuses on action plans (i.e., responses) for when MEGs are exceeded. Actions plans should include, but should not be limited to the following elements:

- Formulating better characterizations of exposures, including identification of the sources and of the contributions from various contaminated media. (More extensive discussion on exposure assessment is provided in Chapter 3.)
- Setting limitations on the lengths of deployments.
- Identifying remedial options.
- Identifying exposed individuals who are at greater risk for adverse effects, triggering one or more of the following actions:

  —Post-deployment follow-up with exposed individuals.
  —Identification of unusually susceptible individuals.
  —Limitations on multiple deployments.
  —Consideration of the possibilities of other exposures contributing to the same health outcomes.
  —Provision of long-term care.

Assessing Aggregate Exposure

As discussed in Chapter 3, aggregate exposure is total exposure to a single chemical by multiple pathways and routes. The paths that chemicals travel to reach the media through which individual exposures occur are referred to as *exposure pathways*. Most pathways are complex. For example, lead added to gasoline (medium 1) is emitted to the air (medium 2) when gasoline is burned. Some of the airborne lead is deposited in soil (medium 3), which is used for growing corn. Some of the lead in soil dissolves in water (medium 4) and moves through the roots of the corn plant, accumulating in the kernels of corn (medium 5), and the corn is fed to dairy cattle, leading some of the lead to be excreted in cows' milk (medium 6). In this scenario, milk is the medium through which humans are exposed to the lead. The lead passed through six media before it reached human beings. To make matters more complex, humans could have been exposed to lead at several other points along the pathway—for example, by breathing the air (medium 2) or coming into contact with the soil (medium 3). *Exposure routes* are the ways that chemicals can move into the body. They include inhalation, ingestion, skin absorption, absorption through the eyes, placental transfer from a pregnant woman to the fetus, and transfer from mother to child through breast-feeding. In many cases, contaminated media (air, soil, or water) can lead to several exposure routes. For example, humans could be exposed to an organic solvent in tap water through drinking the contaminated water or through inhaling chemical vapors during warm showers. All of the exposure possibilities should be considered when assessing human health risks.

During short-term missions and deployments, the route of exposure most likely to be of relevance to deployed personnel is inhalation. During longer-term deployments, deployed personnel might be exposed to low levels of common contaminants through various environmental media. In those longer-term scenarios, personnel could inhale contaminants in air that were volatilized from soil and/or water; ingest contaminated water; and/or experience dermal exposures from bathing or from direct contact with contaminated soils. Assessing risk from each exposure route independently might indicate "low to moderate" risk categories; however, considering these potential exposures in aggregate could indicate more significant risks.

EPA's hazard index (HI) method is an example of a simple aggregate-exposure assessment framework that could be implemented by DOD. EPA defines the HI method as an aggregation of individual hazard quotients (HQs) for each route of exposure. The HQs are ratios of exposures to reference concentrations. For example, the HQ for inhalation is calculated as follows:

$$HQ_{inhalation} = \frac{\text{Exposure Concentration (mg/m}^3\text{)}}{\text{RfC (mg/m}^3\text{)}}. \qquad (5\text{-}4)$$

An oral RfD, a dermal RfD, and/or an inhalation RfC must be defined for each route of concern. HQs for each route of concern can then be aggregated into an HI.

$$HI_{pathway} = HQ_{oral} + HQ_{dermal} + HQ_{inhalation}. \qquad (5\text{-}5)$$

Risk increases with increasing HQs and HIs. Generally, HQs or HIs of less than or equal to 1 are of little concern, whereas HQs or HIs of greater than 1 are of greater concern. This is a simple summary of the EPA procedure. EPA's methodology for aggregate exposure (EPA 1999b, 2001b) should be consulted for more details. For the purpose of assessing aggregate exposures involving MEG chemicals, the subcommittee recommends DOD adapt EPA's method for use with MEGs. For example,

$$HQ_{air,\ water,\ or\ soil} = \text{Exposure Concentration/MEG}; \qquad (5\text{-}6)$$

$$HI = HQ_{air} + HQ_{water} + HQ_{soil}. \qquad (5\text{-}7)$$

Assessing Cumulative Risk

TG-230 points out that because "certain contaminants may have similar adverse effects on the human body, it is necessary to consider the total sum of all similar effects" (USACHPPM 2000a). The document further indicates that in the preliminary threat analysis, when occupational and environmental health (OEH) hazards are identified and prioritized, the effects of exposures to the same or similar chemicals through different media should be considered additive. Algorithms have been adopted by federal agencies to address the problem of exposures to multiple chemicals; however, in RD-230 USACHPPM states that those quantitative approaches are "not well-suited to the overall qualitative/ranking nature of the TG-230 deployment risk assessment approach" (USACHPPM 2002b, p. 5). The subcommittee agrees that conventional algorithms used to assess health risks from multiple chemicals are not useful for assessing mission risks. However, for the purposes of force health protection, those algorithms are appropriate for assessing cumulative risks to the deployed force. The most common procedure is discussed below.

Cumulative exposures involve exposures to multiple chemicals. EPA defines cumulative risk as the likelihood of occurrence of an adverse health effect from exposure to multiple chemicals that have common modes of toxicity from all routes and pathways. The subcommittee agrees with the Army's assumption that the toxicity of a mixture of chemicals that have similar modes of action will be equal to the sum of the weighted dose toxicities of the individual chemicals in the mixture. When assuming "additivity," the methods for combining component data described in EPA's *Supplementary Guidance for Conducting Health Risk Assessment of Chemical Mixtures* (EPA 2000) could be implemented.

The primary method for component-based risk assessments of mixtures of chemicals with similar modes of action is the hazard index (HI), which is derived from dose addition. In the EPA guidance, dose addition is interpreted as simple similar action where the component chemicals act as if they were dilutions or concentrations of each other, differing only in relative toxicity. Dose additivity might not hold for all toxic effects, and the relative toxic potency between chemicals might differ for different types of toxicity or for toxicity by different routes. To reflect those differences, an HI usually is developed for each exposure route of interest and for a single specific toxic effect or for toxicity to a single target organ. A mixture could then be assessed using several HIs, each representing one route and one toxic effect or target organ.

EPA's HI is defined as a weighted sum of the exposure measures for the mixture component chemicals. According to dose addition, the "weight" factor should be a measure of the relative toxic strength. The guidelines formula for the HI is general.

$$\underset{i=1}{\overset{n}{HI_1}} = \Sigma\, E_i/AL_i, \qquad (5\text{-}8)$$

where E_i is the exposure level to chemical i, AL_i is the acceptable level for chemical i, and n is the number of chemicals in the mixture. When an effect-specific HI exceeds 1, potential toxicity is a concern. In practice, EPA usually calculates the HIs by using RfDs or RfCs as the *AL*s. By modifying the formula, DOD can utilize other expressions for exposure and relative toxicity that might be more appropriate for deployment situations.

To apply this HI approach to military situations, the relevant MEGs should be used as the *ALs*.

In practice, the HI method could be applied to chemicals that have similar target-organ effects. However, given the range of data sources on which the current MEGs were based, to begin considering cumulative risk the existing information in TG-230 and RD-230 would have to be reorganized by target organs. That will require going back to the source data in some cases to identify all of the end points considered in addition to the critical effects on which the source guidelines were based. It might be more practical to use a qualitative assessment scheme as the first stage in integrating cumulative risk considerations into the MEG guidance.

Repeated Exposures and Multiple Deployments

Many soldiers will participate in multiple deployments during their military career. It is unlikely that multiple short-term deployments (less than or equal to14 days) involving exposures at levels below the MEGs (but not necessarily the CCEGs) will affect the likelihood of toxicity. The impacts of multiple long-term deployments are more relevant to force health protection.

As described in previous sections, long-term MEGs for noncancer effects were based preferentially on subchronic toxicity values. However, in the many cases where only chronic toxicity values were available, those values were used. When long-term MEGs are based on chronic toxicity data they will be protective for lifetime exposures and will, therefore, also be protective for multiple deployments. When long-term MEGs are based on subchronic toxicity data, they will be protective for up to 7 years (10% of lifetime). Thus, a soldier would need to have more than 7 years of deployment exposures to a chemical at concentrations close to the long-term MEG before any concern would arise regarding the health impacts of those multiple deployment exposures. Furthermore, the UFs applied to the noncancer MEGs provide additional protection.

For the long-term MEGs based on cancer risks, risks from multiple deployments might be viewed as irreversible over time. In a worst-case scenario with multiple exposures near MEG concentrations based on 1×10^{-4} incremental cancer risk, it is unlikely that risks from multiple deployments will contribute to total risk in excess of 1×10^{-3}, which is a target risk used to develop many occupational standards. At that upper-bound risk level, risks to an individual soldier are still low compared with the background cancer risks of 0.25 (or one in four).

The subcommittee was informed that DOD is working to record soldiers' exposures, to the extent feasible. Those records would be available if the need for retrospective analysis arose. For example, when exposures in excess of long-term MEGs occur during a deployment, it might be useful to review the records of exposures from previous deployments. However, there is no indication of an imminent need for prospective analysis of these records for the purpose of monitoring future deployments.

RECOMMENDATIONS

This section summarizes the major recommendations for the development and application of MEGs. The chapter itself should be consulted for more thorough descriptions and for several other recommendations that are more detailed, are more chemical-specific, or are of secondary importance. Overall, the MEGs must be re-evaluated and revised to make them more relevant to force health protection and more consistent with each other. Ideally, USACHPPM will develop a set of principles, guidelines, and procedures for developing MEGs *de novo* from the primary toxicology data. Those procedures would solidify the purpose and goals of the MEGs and would make explicit the risk-management policy decisions that underpin the removal or modification of uncertainty factors used in the existing guidelines set by other agencies. However, the subcommittee realizes the immensity of that undertaking and suggests that, in the interim, revisions be made to improve the quality of the MEGs. To assist in obtaining and managing resources for this effort, DOD should analyze the resources (staff and funding) needed to accomplish the recommendations, prioritize the tasks, and estimate how much time it will need to complete this work.

Near-Term Revisions

- Improve the quality of the MEGs by making revisions that require relatively minimal resources. Specifically, USACHPPM has applied some adjustments to the source guidelines to make them relevant to the deployed population but does not appear to have done so consistently. The following are recommended modifications:

—When using TLVs to derive the 14-day and 1-year MEGs, it is unnecessarily conservative to apply a UF of 10 to account for uncertainty associated with extrapolation from intermittent to continuous exposure.

—All relevant MEGs should include the military adjustment factors for higher ventilation and water-intake rates.

—For the six criteria air pollutants, ensure that the MEGs are appropriate to the military population rather than to susceptible civilian subpopulations.

—Periodically review the guidelines set by other organizations that were used as sources for the MEGs. If those sources have been revised, incorporate the changes into the MEGs. Values reported in HEAST should not be used as bases for MEGs, because they have not been peer-reviewed. Additional exposure guidelines should be consulted, such as the RfCs developed by the State of California's Office of Environmental Health Hazard Assessment.

—Improve the documentation of the existing exposure guidelines by specifying the date of their establishment, the toxicity end points on which they are based, UFs used, and any special considerations in the supporting reference tables. Adjustments to the values should be made on a case-by-case basis.

—Develop short-term soil MEGs for certain contaminants, particularly volatile organic compounds.

—Re-evaluate the approach used to assess dermal exposures to CWAs.

- Establish risk-management framework that focuses on action plans (i.e., responses) for when the MEGs are exceeded. Appropriate actions would include considering risk-management options for reducing or eliminating risks (e.g., using protective gear) and determining the appropriate medical follow-up responsibilities of DOD (e.g., documenting the exposure in medical records, tracking exposed individuals, providing long-term care) when some health risks must be borne.

Mid-Term Revisions

Steps in this category would result in more relevant and internally consistent MEGs that would be less likely to be overly conservative. However, there should be an analysis of the level of effort required for this activity relative to that required for the long-term revisions. The optimal approach to revising the MEGs would be for USACHPPM to consult the original source material (e.g., the critical study selected by EPA for an RfC or cancer unit risk) and perform its own calculations. That would bring more unity to the guidelines. The effort would be time-consuming and would have flaws resulting from out-of-date source materials. However, it would avoid the more time-consuming tasks of literature searches and evaluations of the primary literature while providing a transparent, systematic, and uniform method of applying adjustments to exposure durations, inhalation rates, and water intake rates. It would also standardize the treatment of susceptible subpopulations. In some cases, other agencies could be asked to provide assistance. Discussions with other agencies also might reveal possibilities for accessing professionals already familiar with the assessments who could go back and make recalculations.

Long-Term Revisions

- As discussed, the source material used to derive the MEGs has inherent problems, the primary problem being the obsolescence of many of the values. Thus, simply changing the UFs and other factors will not solve all of the underlying difficulties. However, it is not feasible for USACHPPM to create MEGs entirely *de novo* by beginning with literature searches. All of the agencies developing health-based guidelines struggle with the problem of obsolescence. For example, EPA is beginning a major effort to reinvigorate IRIS. That presents an opportunity to explore partnership arrangements. For example, if one agency were doing a *de novo* assessment on a chemical of interest to the military, it would be relatively easy for that agency to establish one guideline applicable to their interests and another applicable to the military. The word relatively is used because the major effort in assessments is evaluating and interpreting the literature, which would have to be done by the agency as well as the military.
- Aggregate exposure and cumulative risk should be addressed, to the extent feasible, in each stage of revisions to the MEGs.
- USACHPPM should periodically update the list of chemicals for which MEGs have been derived to include chemicals that were omitted in previous reviews (e.g., gasoline) or that have been newly identified as contaminants.

REFERENCES

ATSDR (Agency for Toxic Substances and Disease Registry). 2003. Minimal Risk Levels

(MRLs). Agency for Toxic Substances and Disease Registry, Atlanta, GA. [Online]. Available: www.atsdr.cdc.gov/mrls.html [accessed Dec. 3, 2003].

Bowers, T.S., and J.S. Cohen. 1998. Blood lead slope factor models for adults: Comparisons of observations and predictions. Environ. Health Perspect. 106(Suppl. 6):1569-1576.

Bowers, T.S., B.D. Beck, and H.S. Karam. 1994. Assessing the relationship between environmental lead concentrations and adult blood lead levels. Risk Anal. 14(2):183-189.

EPA (U.S. Environmental Protection Agency). 1996a. Air Quality Criteria for Ozone and Related Photochemical Oxidants. EPA/600/P-93/004cF. National Center for Environmental Assessment, Office of Research and Development, U.S. Environmental Protection Agency, Washington, DC.

EPA (U.S. Environmental Protection Agency). 1996b. Soil Screening Guidance: User's Guide, 2nd Ed. Pub. No. 9355.4-23. EPA540/R-96/018. Office of Solid Waste and Emergency Response, U.S. Environmental Protection Agency, Washington, DC. [Online]. Available: http://www.epa.gov/superfund/resources/soil/index. htm#user [accessed Dec. 5, 2003].

EPA (U.S. Environmental Protection Agency). 1996c. Recommendations of the Technical Review Workgroup for Lead for an Interim Approach to Assessing Risks Associated with Adult Exposures to Lead in Soil. Environmental Criteria and Assessment Office, U.S. Environmental Protection Agency, Research Triangle Park, NC.

EPA (U.S. Environmental Protection Agency). 1996d. Drinking Water Regulations and Health Advisories. EPA822-R-96-001. Office of Water, U.S. Environmental Protection Agency. October 1996.

EPA (U.S. Environmental Protection Agency). 1998. U.S. EPA Region 9 Preliminary Remediation Goal Tables. Office of Solid and Hazardous Waste, U.S. Environmental Protection Agency.

EPA (U.S. Environmental Protection Agency). 1999a. EPA Region 3 Risk-Based Concentration Table. Risk Assessment, Office of Solid Waste and Emergency Response, U.S. Environmental Protection Agency.

EPA (U.S. Environmental Protection Agency). 1999b. Guidance for Performing Aggregate Exposure and Risk Assessments, Draft, February 1, 1999. Office of Pesticide Programs, U.S. Environmental Protection Agency. [Online]. Available: www.epa.gov/scipoly/sap/1999/february/guidance.pdf [accessed Dec. 5, 2003].

EPA (U.S. Environmental Protection Agency). 2000. Supplementary Guidance for Conducting Health Risk Assessment of Chemical Mixtures. EPA/630/R-00/002. Risk Assessment Forum, U.S. Environmental Protection Agency, Washington, DC. [Online]. Available: http://www.epa.gov/iris/backgr-d.htm [accessed Dec. 3, 2003].

EPA (U.S. Environmental Protection Agency). 2001a. Risk Assessment Guidance for Superfund—Volume I: Human Health Evaluation Manual. (Part E, Supplemental Guidance for Dermal Risk Assessment), Interim Review Draft. EPA/540/R/99/005. OSWER 9285.7-02EP. PB99-9633112. Office of Emergency and Remedial Response, U.S. Environmental Protection Agency, Washington, DC. [Online]. Available: http://www.epa.gov/oerrpage/superfund/programs/risk/ragse/ [accessed Dec.3, 2003].

EPA (U.S. Environmental Protection Agency). 2001b. General Principles for Performing Aggregate Exposure and Risk Assessments. Office of Pesticide Programs, U.S. Environmental Protection Agency. [Online]. Available: http://www.epa.gov/pesticides/trac/science/aggregate.pdf [accessed Dec. 3, 2003].

EPA (U.S. Environmental Protection Agency). 2002a. U.S. EPA Region 3 Risk-Based Concentration Table. Risk Assessment, Office of Solid Waste and Emergency Response, U.S. Environmental Protection Agency. [Online]. Available: www.epa.gov/reg3hwmd/risk/ [updated semiannually].

EPA (U.S. Environmental Protection Agency). 2002b. U.S. EPA Region 9 Preliminary Remediation Goal Tables. Office of Solid and Hazardous Waste, U.S. Environmental Protection Agency. [Online]. Available: http://www.epa.gov/region09/waste/sfund/prg/index.htm [updated annually].

EPA (U.S. Environmental Protection Agency). 2002c. Supplemental Guidance for Developing Soil Screening Levels for Superfund Sites. OSWER 9355-4-24. Office of Solid Waste and Emergency Response, U.S. Environmental Protection Agency, Washington, DC. [Online]. Available: http://www.epa.gov/ superfund/ programs/ risk/toolthh.htm [accessed Dec. 3, 2003].

EPA (U.S. Environmental Protection Agency). 2002d. Blood Concentrations of U.S. Adult Females: Summary Statistics from Phases 1 and 2 of the National Health and Nutrition Evaluation Survey (NHANES III). OSWER 9285.7-52. Office of Solid Waste and Emergency Response, U.S. Environmental Protection Agency, Washington, DC. [Online]. Available: www.epa.gov/superfund/ programs/lead/products.htm [accessed Dec. 3, 2003].

EPA (U.S. Environmental Protection Agency). 2002e. 2002 Edition of the Drinking Water Standards and Health Advisories. EPA 822-R-02-038. Office of Water, U.S. Environmental Protection Agency, Washington, DC. [Online]. Available: http://www.epa.gov/waterscience/drinking/ [accessed Dec. 3, 2003].

Gaylor, D.W. 2000. The use of Haber's law in standard setting and risk assessment. Toxicology 149(1):17-19.

Graham, J.A., L. Folinsbee, J.M. Davis, J. Raub, and L.D. Grant. 1999. Critical health issues of criteria air pollutants. Pp. 365-397 in Toxicology of the Lung, 3rd Ed, D.E. Gardner, J.D. Crapo, and R.O. McClellan, eds. Philadelphia, PA: Taylor and Francis.

Henry, C.D. 1985. Heat stress and its effects on illness and injury rates. Mil. Med. 150(6):326-329.

Holmes, K.K., J.H. Shirai, K.Y. Richter, and J.C. Kissel. 1999. Field measurements of dermal soil loading in occupational and recreational activities. Environ. Res. 80(2Pt1):148-157.

Kissel, J.C., K.Y. Richter, and R.A. Fenske. 1996. Field measurement of dermal soil loading attributed to various activities: Implications for exposure assessment. Risk Anal. 16(1):115-125.

Kissel, J.C., J.H. Shirai, K.Y. Richter, and R.A. Fenske. 1998. Investigation of dermal contact with skin in controlled trials. J. Soil Contam. 7(6):737-752.

NRC (National Research Council). 1986. Criteria and Methods of Preparing Emergency Exposure Guidance Level (EEGL), Short-Term Public Emergency Guidance Level (SPEGL), and Continuous Exposure Guidance Level (CEGL) Documents. Washington, DC: National Academy Press.

NRC (National Research Council). 2000. Strategies to Protect the Health of Deployed U.S. Forces: Analytical Framework for Assessing Risks. Washington, DC: National Academy Press.

O'Flaherty, E.J. 1993. Physiologically based models for bone-seeking elements. IV: Kinetics of lead disposition in humans. Toxicol. Appl. Pharmacol. 118(1):16-29.

USACHPPM (U.S. Army Center for Health Promotion and Preventive Medicine). 1999. Derivation of Health-based Environmental Screening Levels for Chemical Warfare

Agents. U.S. Army Center for Health Promotion and Preventive Medicine, Aberdeen Proving Ground, MD, and Oak Ridge National Laboratory, Oak Ridge, TN.

USACHPPM (U.S. Army Center for Health Promotion and Preventive Medicine). 2002a. Chemical Exposure Guidelines for Deployed Military Personnel. Technical Guide 230. U.S. Army Center for Health Promotion and Preventive Medicine. January 2002. [Online]. Available: http://chppm-www.apgea.army.mil/deployment/ [accessed November 25, 2003]

USACHPPM (U.S. Army Center for Health Promotion and Preventive Medicine). 2002b. Chemical Exposure Guidelines for Deployed Military Personnel. A Companion Document to USACHPPM Technical Guide (TG) 230 Chemical Exposure Guidelines for Deployed Military Personnel. Reference Document (RD) 230. U.S. Army Center for Health Promotion and Preventive Medicine January 2002. [Online]. Available: http://chppm-www.apgea.army. mil/deployment/ [accessed November 25, 2003]

U.S. Department of the Army. 1999. Sanitary Control and Surveillance of Field Water Supplies, Draft, May 1999. Technical Bulletin, Medical 577. (as cited in USACHPPM 2002a)

Wester, R.C., H.I. Maibach, L. Sedik, J. Melendres, C.I. Liao, and S. DiZio. 1992. Percutaneous absorption of [14C]chlordane from soil. J. Toxicol. Environ. Health 35(4):269-277.

WHO (World Health Organization). 2000. Air Quality Guidelines for Europe, 2nd Ed. European Series No 91. Copenhagen: World Health Organization.

Appendixes

Appendix A

Errata, Inconsistencies, and Comments on Specific Aspects of TG-248, TG-230, and RD-230

TABLES AND CHARTS

The hazard severity ranking charts presented in TG-248 and TG-230 are not entirely consistent with each other (see Tables 2-2 and 2-3 in Chapter 2 of this document). In general, the table provided in TG-230 (Table 2-2) is more detailed and contains probability ranges that differ somewhat from those that appear in TG-248 (Table 2-3). Moreover, the greater detail provided in the TG-230 table includes probabilistic operational definitions that more clearly distinguish between hazard types than those provided in TG-248. Specifically, TG-230 associates "catastrophic," "critical," "marginal," and "negligible" hazard types with health outcomes involving the development of severe illness in >25%, 10-25%, ≤10%, and 0% of exposed personnel, respectively. Marginal and negligible hazards are also associated with the development of mild illness in >10% and 0-10% of exposed personnel, respectively. These definitions present a minor ambiguity in the distinction between critical and marginal hazards, insofar as both include a 10% level of severe illness among exposed personnel, but that ambiguity can easily be resolved by adjusting the definition of marginal hazard to refer to the threat of severe illness in <10% rather than ≤10% of exposed personnel.

In the hazard probability ranking chart (see Table 2-5 in Chapter 2), the probability ranges used in the table are not properly defined. The notation used should be modified (e.g., change affected-personnel percentages [*P*] as follows: unlikely, $P \leq 10\%$; seldom, $10\% < P \leq 25\%$; occasional, $25\% < P \leq 50\%$; likely, $50\% < P \leq 75\%$; and frequent, $P > 75\%$).

The terminology in Tables 3-3 and 3-4 of TG-248 appears to be inconsistent, because the term "extremely high" is used in Table 3-3, but the term "extreme" is used in Table 3-4. The term used in FM 100-14 is "extremely high."

A risk-assessment summary table is presented in Appendix F of TG-230 (Table F-1). It appears to be useful for working through the steps of the risk-management framework. However, the table is not introduced or discussed in the main body of the document. The subcommittee recommends that the table be moved into the main body and discussed. "Commander's Summary OEH Chemical Risk Assessment" is a possible title. Note that there is a disparity between the "operational risk estimate" in Table F-1 of TG-230 and the risk assessment matrix in Table 3-3 of TG-248. The "risk level" in Table F-1 seems to be equivalent to the "risk estimate" in Table 3-3. Because FM 100-14 uses "risk estimate," "risk level" in Table F-1 should be changed to "risk estimate."

CRITERIA POLLUTANTS

A few specific comments on the text pertaining to criteria pollutants in RD-230 are provided below.

Page 19, ozone. Delete the last sentence. The 8-hour standard is not more "conservative"; it is lower, but the rationale for that is not conservatism. Also, the 1-hour standard is not being revoked in attainment areas. That subject is under litigation.

Page 20, particulate matter. The last paragraph should include a few sentences about particle size and its importance.

Page 20, sulfur dioxide. Delete the sentence about acid rain. This section is about health, so a discussion of the ecological effects of acid rain is too sweeping to be accurate. Also, delete the sentence about visibility. If visibility is to be discussed, it should be part of the section on particulate matter. Sulfur dioxide is a major precursor to secondary particles, but that is only part of the problem of visibility.

Page 20, carbon monoxide. There are 1-hour and 8-hour NAAQS.

Page 21, nitrogen dioxide. Second sentence states "NO_2 can irritate the lungs, cause bronchitis and pneumonia, and lower resistance to respiratory infections." Reference to bronchitis and pneumonia should be deleted and replaced with "changes in pulmonary function," because pulmonary function changes are caused by NO_2, while the other changes have only been associated with respiratory infections and NO_2. Delete the reference to acid rain for the reason listed above. The last sentence refers to the pollutant standard index (PSI), but the PSI is not discussed for the other criteria air pollutants. PSI should be mentioned for all or none. The statement that nitrogen dioxide at 200 ppm is considered by EPA to be "very unhealthy" should be corrected. According to the reference (EPA 64 Fed. Reg. 43530 [1999]), 1-hour levels from 0.65 to 1.24 ppm are considered "very unhealthy," and that term is equated to an air quality index of 201-300. Nitrogen dioxide at 200 ppm is likely to be lethal or close to lethal and not realistically ambient.

Page 38, paragraph "the analyses…", L4. Many cities do not "frequently, if not routinely, exceed the NAAQS." That statement is true for ozone, and one could argue about particulate matter, but it is not true for the other criteria pollutants. It is also a complicated issue, because the NAAQS are actually "design values," not specific numbers as implied by the tables. Regardless, this information is not useful to the RD-230. The sentence should be deleted because it is both incorrect and irrelevant to deployment situations.

RD-230 SOIL SECTION

Page 53. Section 3.4 header is listed as "3.3."

Pages 59-60. Subheadings under Section 3.4.5 are all erroneously listed as 3.4.6.

Page 61. Equation 3-17 title refers to the "Stern" model. It should instead refer to the "Bowers" model.

Page 62. Table RD 3-7 refers to Section 3.2.4 in the rationale box. It should refer to Section 3.4.6.3.

Page 64. The first paragraph cites USEPA 1989b as the Exposure Factors Handbook (EFH). The EFH was issued in 1997 (USEPA 1997c).

Appendix B

Review of Acceptable Cancer Risk Levels

ORGANIZATIONAL USE OF ACCEPTABLE RISK LEVELS

Federal Agencies

In an evaluation of regulated chemicals, Travis and Hattemeyer-Frey (1988) found that 70% of chemical carcinogens have a post-regulatory added lifetime estimated risk greater than 1×10^{-6}, and 30% have a post-regulatory added lifetime estimated risk greater than 1×10^{-4}. They concluded that past regulatory decisions explicitly acknowledge that some risks, in the range of 1×10^{-6} to 1×10^{-3}, are acceptable in modern society. In the case of benzene, high estimates of maximum individual risk (1×10^{-3}) are considered tolerable when the aggregate population risk (extra cancers per year in the population) is insignificant. Similar levels of risk are accepted for emissions from zinc-oxide plants (3×10^{-3}), secondary lead smelters (3×10^{-3}), elemental phosphorus plants (1×10^{-3}), vinylidine chloride facilities (8×10^{-4}), DOE facilities emitting radionuclides (7×10^{-4}), and uranium mill tailings emitting radon (5×10^{-4}) (EPA 48 Fed. Reg. 33112 [1983]; 50 Fed. Reg. 5190 [1985]; 50 Fed. Reg. 32632 [1985]; 51 Fed. Reg. 6382 [1986]; Rodricks et al. 1987; Travis and Hattemeyer-Frey 1988).

Occupational Safety and Health Administration

In deciding on a level of acceptable risk associated with the workplace, the Occupational Safety and Health Administration (OSHA) used an approach similar to that of the U.S. Food and Drug Administration (FDA) by not defining "safe" as the equivalent of risk-free, because many activities considered safe by most people entail some risk of accident or health damage. Workplace activities or exposures are not considered unsafe unless a significant risk of harm exists. In addition, because of the benefits accrued from employment (e.g., income), workers are presumed to be willing to accept higher levels of risk than would someone to whom little or no benefit accrues from accepting risk. Some studies have shown that salary is commensurate with the level of risk inherent in an occupation (Starr 1969; Whipple 1988).

Supreme Court action (*Industrial Union Department, AFL-CIO v. American Petroleum Institute et al.* 448 U.S. 607 [1980]) was instrumental in defining acceptable occupational risk for OSHA. The court suggests that significant occupational risk be determined by comparing the risk in question with other common occupational risks levels. The court suggested that an occupational lifetime cancer risk of 1×10^{-3} is significant. On the basis of actuarial data from 1984, the average lifetime (i.e., 45 years) risk of work-related death in a private company with 11 or more employees was 2.9 per 1,000. Risks in high-hazard occupations, like mining, range between 7.6 and 18.6 per 1,000, and risks in low-hazard workplaces, like the service industry, range between 0.9 and 1.8 per 1,000 (Cotter 1986; Rodricks et al. 1987). These rates are measured, not estimated risks, and show little variation from year to year.

OSHA as well as the U.S. Environmental Protection Agency (EPA) and the Nuclear Regulatory Commission have used those lifetime risk values as benchmarks to develop occupational acceptable risk levels. In developing radiation protection guidelines, EPA selected 3×10^{-3} because it was comparable to the working lifetime risk of accidental death in the least hazardous occupations (EPA 46 Fed. Reg. 7836 [1981]; Rodricks et al. 1987). The Nuclear Regulatory Commission stated that the average annual mortality rate in "safe industries" does not exceed 1×10^{-4}, which translates to a worker 45-year lifetime risk of approximately 4×10^{-3}. Like EPA, the commission proposed occupational standards on the basis of the assumption that worker risks due to radiation are acceptable if kept at or below the "safe industry" risk level (Nuclear Regulatory Commission 1986; Rodricks et al. 1987).

For other workplace carcinogens, OSHA has not regulated below 1×10^{-3}, largely because of technical feasibility. Residual lifetime occupational risks associated with permissible exposure levels (PELs) revised in the

1980s range from a low of 0.2 per 1,000 to 6 per 1,000 for ethylene dibromide and 39 per 1,000 for acrylonitrile (Table B-1). As with other federal agencies, OSHA has not claimed that those health risks are insignificant, but rather that they are "acceptable" (Rodricks et al. 1987).

National Institute for Occupational Safety and Health

Unlike OSHA, the National Institute for Occupational Safety and Health (NIOSH) is not a regulatory agency, but it is engaged in research on and interpretation of occupational health and safety issues. Its policy on acceptable cancer risk is qualitative, as first stated in 1976, calling for "no detectable exposure levels for proven carcinogens" (Fairchild 1976; NIOSH 2002). That statement represents a zero-risk policy similar to FDA's Delaney Clause of 1958 and the treatment of Category I potential carcinogens in the generic OSHA rulemaking on carcinogens (29 CFR 1990). Under the policy, when carcinogen thresholds that would protect 100% of the population have not been identified, nonquantitative recommended exposure levels (RELs) labeled "lowest feasible concentrations" (LFCs) are adopted for most carcinogens. The few quantitative RELs for carcinogens (e.g., for asbestos, formaldehyde, benzene, and ethylene oxide) were set based on limits of detection or technological feasibility. NIOSH also adopted several quantitative RELs from OSHA's 1989 PEL update (implying an acceptable risk limit consistent with that used by OSHA for category II potential carcinogens).

TABLE B-1 Risk Considered Acceptable within OSHA Permissible Exposure Limits

Chemical	Residual Risk at PELs
Arsenic	8×10^{-3}
Ethylene oxide	$1\text{-}2 \times 10^{-3}$
Ethylene dibromide	$0.2\text{-}6 \times 10^{-3}$
Benzene	$5\text{-}16 \times 10^{-3}$
Acrylonitrile	39×10^{-3}
Dibromochloropropane	2×10^{-3}
Asbestos	6.7×10^{-3}

Source: Adapted from Rodricks et al. 1987.

In the 20 years since NIOSH first set its policy, science and risk-assessment techniques and management techniques have advanced to the point that NIOSH has adopted a more inclusive attitude. NIOSH projects both no-effect exposure levels for chemical or physical agents as well exposure levels at which residual risks for all workplace hazards, including carcinogens, might be present. This approach is consistent with the 1970 OSH Act (P.L. 91-596 Section 20(a)(3)) that charged NIOSH to "describe exposure levels that are safe for various periods of employment, including but not limited to the exposure levels at which no employee will suffer impaired health or functional capacities or diminished life expectancy as a result of his work experience." However, no single acceptable risk level or range of values has been set forth in policy to date.

U.S. Environmental Protection Agency

EPA has been at the forefront of the issue of acceptable risk in virtually all of its programmatic areas, primarily as the result of court challenges to its regulations. In response to the 1987 Section 112 Clean Air Act decision (*Natural Resources Defense Council v. U.S. Environmental Protection Agency* 824 F. 2nd 1146 [1987]), EPA decided it would base its regulatory decisions on quantitative risk assessments using the general policy that a lifetime added cancer risk for the most exposed person of 1 in 10,000 (1×10^{-4}) might constitute acceptable risk and that the margin of safety required by statute and reinforced by the court should reduce the risk for the greatest number of persons to an added lifetime risk of no more than 1 in 1 million (1×10^{-6}). However, EPA (along with the courts) has not viewed "safe" as the equivalent of risk-free and has determined that standards should protect against significant public health risks (EPA 49 Fed. Reg. 8386 [1984]; Rodricks et al. 1987; *Industrial Union Department, AFL-CIO v. American Petroleum Institute et al.* 448 U.S. 607 [1980]). EPA has repeatedly rejected the opinion that it can establish a universal (i.e., brightline) acceptable risk that should never be exceeded under any circumstances, and they maintain that guidance provided under one statute might have little relevance to others because of differing program goals. In practical terms, EPA almost never regulates at a theoretical risk below 1×10^{-6} (*de minimis*) and almost always regulates at a theoretical risk below 1×10^{-4} (*de manifestis*).

U.S. Food and Drug Administration

The U. S. Food and Drug Administration (FDA) was the first federal agency known to have encountered the issue of acceptable risk with the adoption of the Delaney Clause in 1958. The clause states that no additive that is deliberately included in food products during or after processing is allowed to be in use if it causes cancer in animals (Federal Food, Drug, and Cosmetic Act Sec. 409.[348](c)(3)(A); Sec. 512.[360b](d)(1)(I); and Sec. 721.[379e](b)(5)(B)). At the time, this zero-risk policy was based on the presumption that no safe human exposures to animal (nonthreshold) carcinogens could be identified the way that safe levels of compounds acting through threshold mechanisms could (NRC 1994).

In 1973, FDA employed quantitative risk assessment using the Mantel-Bryan methodology for estimating cancer risks (Mantel and Bryan 1961; Rodricks et al. 1987) and defined the risks associated with the residues of carcinogenic drugs used in food animals. The acceptable risk level proposed at the time was 1×10^{-8}, as suggested by Mantel and Bryan (Mantel and Bryan 1961; FDA 38 Fed. Reg. 19226 [1973]; Rodricks et al. 1987). The risk-assessment methodology was later modified to a linear-proportional form, and the acceptable risk level changed to 1×10^{-6} in response to public comment, the idea of risk assessment having become firmly entrenched by 1979 (FDA 44 Fed. Reg. 17070 [1979]; Rodricks et al. 1987). FDA has since employed this technique for other substances regulated under the federal Food, Drug, and Cosmetic (FD&C) Act. D&C Green No. 5 was listed with an estimated upper limit risk of 3×10^{-7}, and lead acetate also received approval as a colorant with a lifetime added risk of between 5×10^{-6} and 1.9×10^{-7} (FDA 45 Fed. Reg. 72112 [1980]; 47 Fed. Reg. 24278 [1982]; Rodricks et al. 1987). In neither instance was 1×10^{-6} declared an agency-significant risk criterion, but the above risks were considered insignificant in terms of the public health as applied to food additives.

The same risk-assessment tools have been used to address the zero-risk requirements of the Delaney Clause. In those cases, FDA has interpreted "safe" in the context of food law to be defined as "reasonable [but not absolute] certainty of no harm," although at the same time, the benefits of an additive cannot be considered in its decisions. The FDA position is that a carcinogen is considered safe as long as exposure to it is restricted to levels posing insignificant risks. Insignificant risk has been defined as 1×10^{-6} or less in several agency decisions.

Even though 1×10^{-6} has been used as the *de facto* brightline acceptable risk level in several cases, FDA has also found that cancer risks above 1×10^{-6} are acceptable for certain classes of contaminants (e.g., PCBs, dioxins, and aflatoxins) given the technical difficulties and costs associated with reducing exposure. In the case of fish contaminated with PCBs, FDA considered that an estimated risk of 1×10^{-4} provided adequate protection to the public health on the basis of their PCB residue tolerance of 2 ppm. This conclusion considered the ubiquity of the contaminant and the presumed benefit from consuming fish proteins. FDA stopped short of labeling estimated risks greater than 1×10^{-6} as insignificant (Rodricks et al. 1987).

Other Authoritative Sources

American Conference of Governmental Industrial Hygienists

The American Conference of Governmental Industrial Hygienists (ACGIH), like NIOSH, develops recommended occupational exposure limits to protect workers from injury. Their Threshold Limit Values (TLVs) are not enforceable because ACGIH is a nongovernmental organization. The TLVs are based on threshold toxicity events and include a designation for the weight-of-evidence for carcinogenicity and other forms of toxicity similar to that used by IARC and EPA. ACGIH has not to date identified an acceptable risk for any carcinogen in the workplace, stating instead that exposure to carcinogens must be kept to a minimum (ACGIH 2001). Workers exposed to known human carcinogens that do not have a TLV should be properly equipped to eliminate exposures to those carcinogens to the fullest extent possible. For known human carcinogens that have a TLV or suspected human carcinogens, worker exposures by all routes should be carefully controlled to levels as low as possible below the TLV.

National Commission on Radiologic Protection

The National Commission on Radiologic Protection (NRCP) proposed an average cancer risk of 1×10^{-5} per year (not per worker lifetime or full lifetime) for members of the public exposed continuously or repeatedly to radiation sources other than medical and background. This translates to a lifetime cancer risk of 7×10^{-4} and may be considered the *de manifestis* risk. A negligible 1% of that risk (7×10^{-6}) was suggested by Schiager et al. (1986).

World Health Organization

In developing drinking water guidelines, the World Health Organization (WHO) employed an acceptable risk level of 1×10^{-5} (WHO 1996; Fiori and Meyerhoff 2002) but did not indicate how that level of health protection was selected. For its air quality guidelines, WHO provides only qualitative guidance by specifying that the "acceptability of the risk and, therefore, the standards selected, depends on the expected incidence and severity of the potential effects, the size of the population at risk, the perception of the related risk and the degree of scientific uncertainty that the effects will occur at a specific level of air pollution" (WHO 2000).

REFERENCES

ACGIH (American Conference of Governmental Industrial Hygienists). 2001. Threshold Limit Values for Chemical Substances and Physical Agents and Biological Exposure Indices, 7th Ed. American Conference of Governmental Industrial Hygienists, Cincinnati, OH.

Cotter, D.M. 1986. Work-related deaths in 1984: BLS survey findings. Research summaries. Monthly Labor Review (May):42-44.

Fairchild, E.J., II. 1976. Guidelines for a NIOSH policy on occupational carcinogenesis. Ann. N. Y. Acad. Sci. 271:200-207.

Fiori, J.M., and R.D. Meyerhoff. 2002. Extending the threshold of regulation concept: De minimis limits for carcinogens and mutagens. Regul. Toxicol. Pharmacol. 35(2 Pt 1):209-216.

Mantel, N., and W.R. Bryan. 1961. A safety testing of carcinogenic agents. J. Natl. Cancer Inst. 27:455-470.

NIOSH (National Institute for Occupational Safety and Health). 2002. NIOSH Pocket Guide to Chemical Hazards, Appendix A. NIOSH Potential Occupational Carcinogens. National Institute of Occupational Safety and Health, Centers for Disease Control, U.S. Dept. of Health and Human Services, Cincinnati, OH. [Online]. Available: http://www.cdc.gov/niosh/npg/nengapdx.html#a [accessed Dec. 5, 2003].

NRC (National Research Council). 1986. Criteria and Methods of Preparing Emergency Exposure Guidance Level (EEGL), Short-Term Public Emergency Guidance Level (SPEGL), and Continuous Exposure Guidance Level (CEGL) Documents. Washington, DC: National Academy Press.

NRC (National Research Council). 1994. Science and Judgment in Risk Assessment. Washington, DC: National Academy Press.

Rodricks, J.V., S.M. Brett, and G.C. Wrenn. 1987. Significant risk decisions in federal regulatory agencies. Regul. Toxicol. Pharmacol. 7(3):307-320.

Schiager, K.M., W.J. Bair, M.W. Carter, A.P. Hull, and J.E. Till. 1986. De minimis environmental radiation levels: Concepts and consequences. Health Phys. 50(5):569-579.

Starr, C. 1969. Social benefit versus technological risk. Science 165(899):1232-1238.

Travis, C.C., and H. Hattemer-Frey. 1988. Determining an acceptable level of risk. Environ. Sci. Technol. 22(8):873-876.

Whipple, C. 1988. Acceptable risk. Pp. 157-170 in Carcinogen Risk Assessment, C.C. Travis, ed. New York, NY: Plenum Press.

WHO (World Health Organization). 1996. Guidelines for Dinking Water Quality, 2nd Ed. Geneva: World Health Organization.

WHO (World Health Organization). 2000. Air Quality Guidelines for Europe, 2nd Ed. European Series No 91. Copenhagen: World Health Organization.

Appendix C

Example Use of Probits for Developing Chemical Casualty Estimating Guidelines

INTRODUCTION

Chemical casualty estimating guidelines (CCEGs) were introduced by the subcommittee for the purpose of better informing commanders of potential decrements in troop strength that might jeopardize the ability of troops to successfully complete missions. The CCEGs will provide data sets that can be used quantitatively to estimate the nature and extent of impacts on mission performance. Specifically, CCEGs are tools by which the severity of adverse outcomes to mission accomplishment can be estimated from chemical concentrations. Generally they will be set for atmospheric compounds whose toxic potency is quite high. The application of these tools will generate specific response rates (e.g., 25%, 50%) for defined concentrations of chemicals in the breathing zone of military personnel.

The health impacts of chemical agents generally form a continuum from mild physical or sensory alterations—such as mild skin irritation—that pose distractions but are easily accommodated, to impairment of vision and balance that might limit effective use of battlefield equipment, to central nervous system (CNS) depression that would limit necessary cognitive functions, to asphyxiation or serious organ damage and failure leading to death. The medical outcome depends directly on the delivered dose and the inherent toxicity of the chemical. For airborne substances, exposures are characterized by the concentration(s) in air at the breathing zone and the duration of contact.

For field commanders to make informed choices between one course of action and another, arguably less dangerous one (within the context of an assortment of many types of risks), appropriate comparisons must be made. Two types of information in particular are needed to make such comparisons of estimated impacts on troop viability and vulnerability: (1) the severity of the immediate medical consequences during the course of the mission, and (2) the likely number of troops affected in the exposure scenario envisioned during the course of the specific mission.

OBJECTIVE

The purpose of the exercise reported herein is to explore the feasibility of using an approach that relies in part on probit analysis to describe, for three levels of severity (mild, moderate, and severe), the expected exposure-incidence response for the reasonably healthy young adults that comprise the deployed population. As discussed in Chapter 4, this is not a definitive protocol for how to develop CCEGs. Rather, it is an example of one possible approach. The subcommittee recommends that the U.S. Department of Defense (DOD) develop guidance for establishing CCEGs and have that methodology peer-reviewed before application (see Chapter 4).

This analysis focuses on inhalation as the dominant pathway of exposure for military personnel. Unlike military exposure guidelines (MEGs) and other methods to identify levels of exposure that are unlikely to cause injury, the CCEGs (as envisioned here) are media-specific chemical concentrations expected to cause health impairments sufficient to reduce unit strength and therefore pose what the Army calls a medical threat. They are designed to evaluate course-of-action options that are expected to involve chemical exposures. Combat situations can result in human casualties that, although undesired, must nevertheless be tolerated to achieve military objectives. With that in mind, CCEGs allow commanders to weigh chemical risks against other operational risks and decide which chemical risks should be avoided and which must be borne for the sake of the mission.

Note that the approach described herein is intended to produce information about potential health impacts in a form that would allow field commanders to compare the impacts on the achievability of mission objectives from an assortment of chemical and nonchemical hazards that could degrade mission effectiveness. That is accomplished by using an approach that estimates the percentage of troops likely to be incapacitated (and the nature and duration of that incapacitation) by exposures to toxic agents.

Such output is comparable to outputs from other processes that estimate, for example, casualties from enemy fire or from weather conditions that might disable mechanized equipment.

APPROACH AND ORGANIZATION

The probit analysis the subcommittee envisions is predicated on having available incidence data for acute toxicity (i.e., for effects that materialize within minutes to several days following initial exposure). Those data must be reliable to provide useful guidance; peer-review is one major means of achieving the desired level of reliability and predictability.

To be most useful, the toxicity data should span three levels of severity:

- **Mild pathological responses.** Most commonly sensory discomfort and irritation and some mild non-sensory effects observed in groups containing a range of normally distributed susceptibilities that would also be found in populations of healthy, young adults.
- **Moderate pathological responses.** Temporarily debilitating systemic dysfunctions in groups containing a range of normally distributed susceptibilities that would be found in healthy, young adults.
- **Severe pathological responses.** Reversible or irreversible damage to organ functions that is incapacitating, life-threatening, or actually lethal observed in groups comparable to healthy, young adults.

This scheme resembles the graded acute exposure guideline levels (AEGLs) of the U.S. Environmental Protection Agency (NRC 2001) in several respects, and similar classifications might be adopted for CCEGs.

The data should be subjected to some form of weight-of-evidence analysis in which the quality of the data are examined critically and the degree of consistency and concordance is evaluated closely. That process should include some decision rules for determining the relative value of and reliance on primarily human and secondarily animal data, and vice versa.

Several compounds were identified as prospective candidates for this feasibility exercise. The compounds are divided into two groups: (1) those for which AEGLs have been published (NRC 2000, 2002, 2003), and (2) a sampling of compounds identified in U.S. Army Center for Health Promotion and Preventive Medicine's Technical Guide 230 (TG-230), but not included among the AEGLs, that would likely have some applicable incidence data. The compounds in each group are listed in Table C-1.

TABLE C-1 Candidate Compounds

Compounds in TG-230 with AEGLs	Compounds in TG-230 with No AEGLs
Aniline	Acrolein
Arsine	Ammonia
Diborane	Benzene
Dimethylhydrazine	Carbon tetrachloride
Hydrogen cyanide	Carbon monoxide
Hydrogen sulfide[a]	Ethylene oxide
Methyl isocyanate	Formaldehyde
Monomethylhydrazine	Hydrazine
Nerve agents: GA, GB, GD, GF, VX	Methyl bromide
Phosgene	Toluene diisocyanate
Propylene glycol dinitrate	
Sulfur mustard	
1,1,1,2-tetrafluoroethane (HFC-134a)	
1,1-dichloro-1-fluoroethane (HCFC-141b)	

[a]A draft AEGLs document is available, but has not yet been finalized.

From among the candidate compounds, seven substances were selected, and their acute toxicity data was analyzed to obtain data relevant to the plotting of incidence on the basis of the three categories: mild, moderate, and severe. The dose unit selected for this exercise, a reflection of exposure via the inhalation pathway, is parts per million-hour (ppm-hour) (and ppm-minute [min] for acrolein), which is the product of concentration and time ($C \times t$). To the extent possible, the same unit was used for all plots to simplify comparisons. Probits were plotted for the log-dose versus the log-percent response between 2% and 98% (see data sheets and graphs at the end of this appendix; note that the probit scale is on the left side of the plot). Each curve was derived by plotting the values for the data. Alternatively, however, probit values could also be calculated against log-dose. Indeed, calculating at least two probit values for each curve offers the advantage of being able to estimate any point on the curve in order to estimate the expected consequences for specific log-doses. In either case, the plots can be computerized to facilitate field comparisons of medical consequences from exposures to toxic agents.

Doses were obtained from data derived from observations of either humans or laboratory animals. For simplicity, uncertainty factors were not used in the analysis. However, when the selected doses were obtained from data from laboratory animals, consideration was given to adjusting the inhalation doses for differences in inhalation rates and metabolic rates between humans and the test species. When adjusting the doses, consideration was given to the application of a methodology that was proposed by EPA (1994) in its guidelines for the derivation of reference concentrations (RfC) for substances present in ambient air. The subcommittee authoring this report chose its own set of relatively simple decision rules for these examples. Those rules were applied to achieve dose-equivalency for inhaled substances when using probit analysis to estimate the number of potential casualties from short-term exposures to toxic agents. Ultimately, DOD would need to develop its own approach and have it peer-reviewed. The decision rules used here apply solely to situations in which human doses are estimated from laboratory animals. They include the following:

- If a substance causes local toxicity (e.g., skin or lung irritation), the inhaled concentration for humans is set equivalent to that obtained from the data in laboratory animals. This was applied to acrolein and sarin ("mild").
- If the substance causes systemic toxicity (e.g., CNS depression) and the effect is caused by the parent substance or stable metabolites (i.e., half-life measured in hours or more), the inhaled concentration for humans is set equivalent to that obtained from the data in laboratory animals. This was applied to aniline, hydrogen cyanide, and sarin ("severe").
- If the substance causes systemic toxicity (e.g., liver injury) and the effect is caused by the reactive parent substance or highly reactive metabolite(s) (i.e., half-life measured in minutes), the inhaled concentration for humans is estimated by adjusting the concentration obtained from the data in laboratory animals in accordance with the body weight of the species to the -0.75 power, because there exists considerable evidence for the validity of such a procedure (Clewell et al. 2002; NRC 2001). This was applied to dimethylhydrazine and propylene glycol dinitrate.

The results of this process were estimated doses for humans that were considered equipotent in their toxic severity.

FINDINGS

The compounds evaluated in detail are aniline, 1,1- and 1,2-dimethylhydrazine, hydrogen cyanide, propylene glycol dinitrate, acrolein, the chemical warfare agent sarin, and hydrogen sulfide. AEGLs values were available for all of these compounds except acrolein. The relevant information for each compound is described at the end of this appendix on two pages: (1) a fact sheet with a summary of the relevant data for the subject compound and (2) the actual plot of the data.

These compounds span a range of chemical classes and acute toxicological manifestations. However, this is merely a small sample that was useful for an initial feasibility study; the exercise would need to be expanded to draw any firm, generalized conclusions about the validity of this means of data analysis and visualization.

With the exception of acrolein, the interpretations of the raw toxicity data on which the chemical plots were based were adopted directly from the AEGL documents (NRC 2000, 2002, 2003; EPA 2002). Reliance on those documents provides a strong element of peer-review that should minimize controversy about data selection and application. Note also that, in many cases, the dose-response relationships are bounded by only a few points and that some points require inference from the range of toxicological information available for a substance. Ultimately, however, the plots enable identification of all values along each curve.

The results indicate that, for the compounds examined, it is generally possible to obtain estimates of toxicological impacts within each of the three categorical groups of severity in terms of the fraction of a group impacted. The exception was dimethylhydrazine. There were no data on that chemical suitable to estimate the frequency of dose-dependent, mild adverse consequences. Also, the information on acrolein is somewhat different from that on the other compounds. For both mild and moderate impacts, the irritation effects are more time-dependent than they are for the other compounds; but for severe outcomes, the impact can be scaled by concentration, keeping time fixed.

Aniline and hydrogen cyanide had the most directly applicable data sets; the data sets for the other compounds were less robust for the purpose of this report. This leads to perhaps the most vital observation: this approach, regardless of its desirability, might be impractical for several reasons. First, the number of compounds having toxicity data in the form required for the approach to be functional appears to be very limited. Although many studies describe changes in pathological severity with increasing doses, they report far less often on the incidence rates in members of the groups observed. That is particularly true for human studies. Indeed, many human studies are merely case reports, and they have the added limitation of having either no exposure data or data that are highly imprecise.

Available animal studies of acute toxicity also have major limitations. Although most compounds have been tested for lethality (lethal dose in 50% of subjects [LD_{50}] and lethal concentration in 50% of subjects [LC_{50}]), few have been tested for less severe effects. Only longer-duration studies report on sublethal effects and the frequency of responders. Another major limitation is the lack of statistical power. Studies deemed most toxicologically significant often have been performed in dogs or monkeys. Such studies frequently use only three to five experimental subjects per group. Thus, the points on the subcommittee's probit plots appear to be more precise than they actually are. This limitation precluded the calculation of standard deviations around data points or confidence intervals around curves.

Table C-2 provides an example of how CCEGs could be used to estimate impacts on troop strength. For the seven example chemicals, the concentrations estimated to severely affect 15%, 30%, 40%, and 50% of the unit are tabulated to correspond to the affiliated operational risk-management (ORM) risk level and unit status, assuming that the fraction F of the unit exposed is $F = 1.0$ (i.e., that 100% of the unit is exposed). A measured or modeled concentration for a given chemical could be compared with the values in the table to estimate the potential impact on missions. (See Chapter 4 and Appendix E for how to estimate mission impacts for cases in which only a percentage of the unit is exposed. See Appendix E for how to estimate impacts from exposures to multiple chemicals.)

CONCLUSIONS

Military field commanders need reliable estimates of the nature and magnitude of health impairment resulting from toxic exposures to agents that could be encountered during missions that have military objectives. This process seems comparable to the estimation of battle casualties when planning missions with combat roles.

Estimating toxic effects that might impair the performance of deployed units during missions is theoretically possible by evaluating the limited number of acutely toxic agents that could be encountered in some battlefield conditions. However, the data needed to perform those evaluations and to obtain reliable estimates easy to use in the field (e.g., graphic representations) appear to be available for only some of the substances of interest to the military. Furthermore, for those compounds for which estimates are feasible and graphic displays are possible, the information should be applied with some understanding of the strengths and limitations of the data, and caution should be exercised to avoid placing too much confidence in the seeming precision of numerical values.

TABLE C-2 Sample CCEGs for Seven Chemicals for "Severe" Response[a]

	Approximate Concentration in Breathing Zone (ppm-hour)			
Chemical	C_{15}	C_{30}	C_{40}	C_{50}
Aniline	1,400	1,600	1,800	1,850
1,1-Dimethylhydrazine	250	540	800	1,400
Hydrogen sulfide	640	680	700	710
Hydrogen cyanide	95	115	130	140
Propylene glycol dinitrate	40	70	90	120
Acrolein	34	40	45	48
Sarin	10	30	52	90
Evaluation	Degree of Medical Threat			
% of unit severely affected, P^*	15%	30%	40%	50%
Unit troop strength[a]	85%	70%	60%	50%
ORM risk level[a]	Low	Moderate	High	Extremely High
Unit status[a]	Green	Amber	Red	Black

[a]Assumes 100% of the unit is exposed.
Abbreviations: C_{15}, concentration estimated to effect 15% of the unit; C_{30}, concentration estimated to effect 30% of the unit; C_{40}, concentration estimated to effect 40% of the unit; C_{50}, concentration estimated to effect 50% of the unit.

Unit Status
Black: Unit requires reconstitution. Unit below 50% strength.
Red: Combat ineffective. Unit at 50-69% strength.
Amber: Mission capable, with minor deficiencies. Unit at 70-84% strength.
Green: Mission capable. Unit at 85% strength or better.

FACT SHEETS AND PROBIT PLOTS[1]

Acrolein	
Mild Effects—irritation of eyes, nose, and throat	
NOAEL (human) = 0.1 ppm for 8 hour	
Data (human) from Stevens et al. 1961:	
0.5 ppm-5 min	25% response
? ppm-min	50% response (estimated)
0.5 ppm-12 min	90% response
Mode of toxic action: local tissue damage on immediate contact, with increasing time of contact	
Allometric scaling: not applicable	
Uncertainty factors: none	
Data plotted for humans:	
2.5 ppm-min	25% response
3.8 ppm-min	50% response (estimated)
6.0 ppm-min	75% response
Moderate Effects—severe irritation of eyes, nose, and throat; respiratory distress	
Data (human) from Sims and Pattle 1957:	
1.2 ppm-5 min	25% response
2.5 ppm-5 min	50% response (estimated)
8 ppm-10 min	75% response
Mode of toxic action: local tissue damage on immediate contact, with increasing time of contact	
Allometric scaling: not applicable	
Uncertainty factors: none	
Data plotted for humans:	
6.0 ppm-min	25% response
3.8 ppm-min	50% response
? ppm-min	75% response
Severe Effects—respiratory failure to mortality	
Data (rat) from Smyth 1956	
8 ppm-4 hour	8% response
10 ppm-4 hour	25% response (estimated)
12 ppm-4 hour	50% response (estimated)
14 ppm-4 hour	75% response (estimated)
	(Continued)

[1]The data used and values reported herein are provided for illustrative purposes only.

Acrolein (continued)
Mode of toxic action: local tissue damage on immediate contact, with increasing time of contact Allometric scaling: 1:1 for rat:human Uncertainty factors: none Data plotted for humans: 32 ppm-hour 8% response 40 ppm-hour 25% response 48 ppm-hour 50% response 56 ppm-hour 75% response
Mode of Action: tissue damage on immediate contact, cumulative with time of contact
Dose-duration relationship: linear, $C^1 \times t = k$ (?)
Delayed sequellae: decrements in respiratory function
For comparison: OSHA PEL = 0.1 ppm-8 hour OSHA STEL = 0.3 ppm-15 min

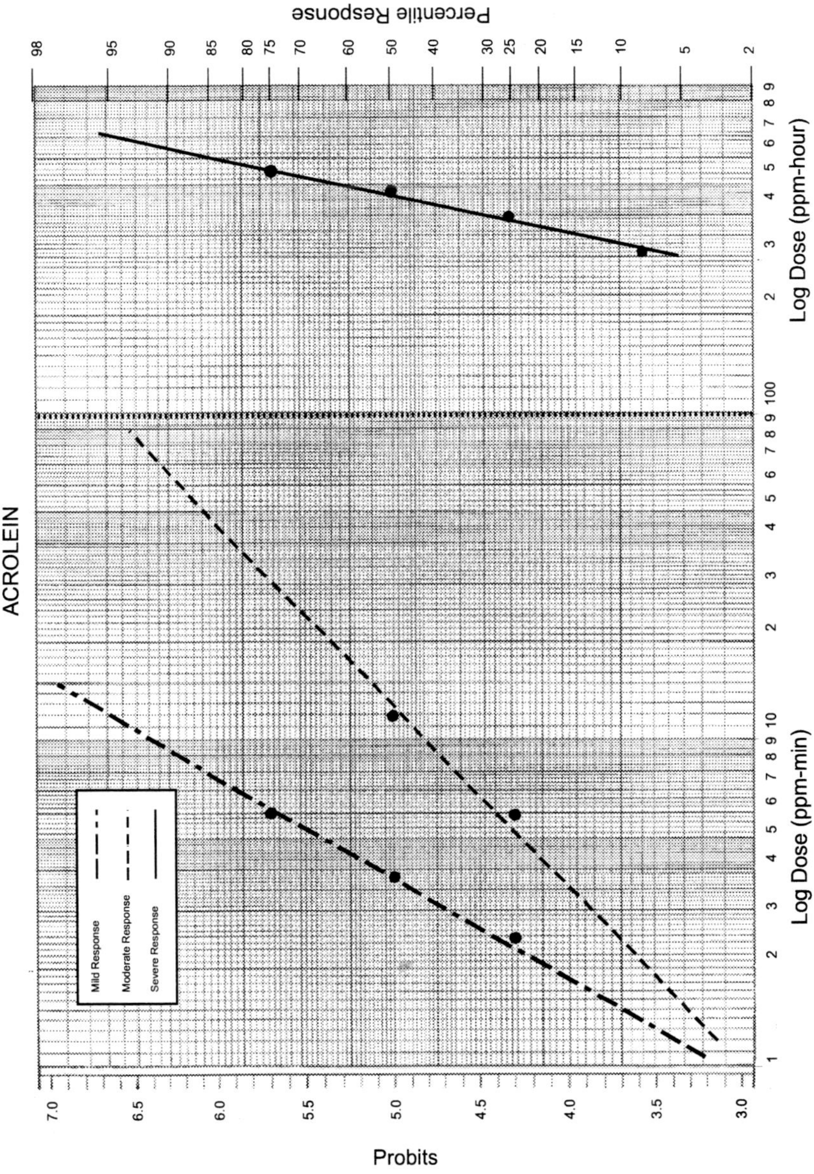
ACROLEIN
Probits
Percentile Response
Log Dose (ppm-min)
Log Dose (ppm-hour)
Mild Response
Moderate Response
Severe Response

Aniline

Mild Effects—clinical cyanosis without hypoxia

NOAEL (human) = 5% MetHb

Data (rats) from Kim and Carlson 1986:

720 ppm-hour	25% response (estimated from database)
800 ppm-hour	50% response (22-23% MetHb)
880 ppm-hour	75% response (estimated from database)

Mode of toxic action: stable metabolite causing MetHb formation

Allometric scaling: 1:1 for rat:human

Uncertainty factors: none

Data plotted for humans:

720 ppm-hour	25% response
800 ppm-hour	50% response
880 ppm-hour	75% response

Source: NRC 2000

Moderate Effects—fatigue, lethargy, dyspnea, headache

Data (rats) from Kim and Carlson 1986:

1,080 ppm-hour	25% response (estimated)
1,200 ppm-hour	50% response (41-42% MetHb)
1,320 ppm-hour	75% response (estimated)

Mode of toxic action: stable metabolite causing MetHb formation

Allometric scaling: 1:1 for rat:human

Uncertainty factors: none

Data plotted for humans:

1,080 ppm-hour	25% response
1,200 ppm-hour	50% response
1,320 ppm-hour	75% response

Source: NRC 2000

Severe Effects—severe hypoxia to asphyxiation to death

Data (rat) from Bodansky 1951, Kiese 1974, and Seger 1992:

1,436 ppm- hour	10% response
1,600 ppm- hour	20% response
1,812 ppm- hour	40% response (70-90% MetHb)
2,120 ppm-hour	70% response

Mode of toxic action: stable metabolite causing MetHb formation

Allometric scaling: 1:1 for rat:human

Uncertainty factors: none

(Continued)

Aniline (continued)
Data plotted for humans: 1,436 ppm-hour 10% response 1,600 ppm-hour 20% response 1,812 ppm-hour 40% response 2,120 ppm-hour 70% response Source: NRC 2000
Mode of Action: MetHb formation at all levels
Dose-duration relationship: linear, $C^1 \times t = k$
Delayed sequellae: possible human carcinogen
For comparison: AEGL-1 = 8 ppm-hour; 1 ppm-8 hour (includes UF of 10 for children) AEGL-2 = 12 ppm-hour; 1.5 ppm-8 hour (includes UF of 10 for children) AEGL-3 = 20 ppm-hour; 2.5 ppm-8 hour (includes UF of 10 for children)

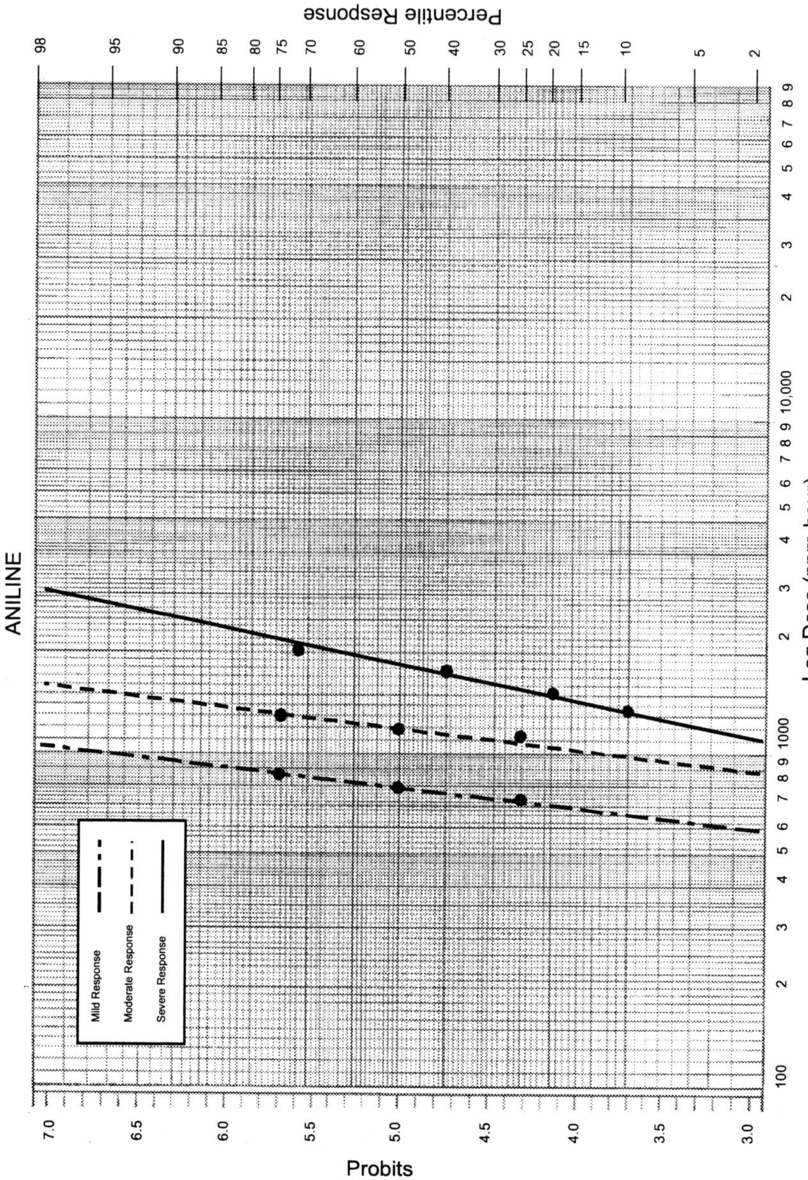
ANILINE
Percentile Response
Probits
Log Dose (ppm-hour)
Mild Response
Moderate Response
Severe Response

1,1- and 1,2- Dimethylhydrazine
Mild Effects—slight tremors of extremities
NOAEL (human): insufficient data Data: insufficient Mode of toxic action: reactive metabolite causing uncharacterized central nervous system effects Allometric scaling: not applicable Uncertainty factors: none Data plotted for humans: insufficient data Source: NRC 2000
Moderate Effects—muscle fasciculations, tremors, vomiting
Data (dog) from Weeks et al. 1963: 80-120 ppm-hour 33% response 200-250 ppm-hour 66% response Mode of toxic action: reactive metabolite causing uncharacterized central nervous system effects Allometric scaling: (body weight)$^{-0.75}$ Uncertainty factors: none Data plotted for humans: 150 ppm-hour 33% response 390 ppm-hour 66% response Source: NRC 2000
Severe Effects—convulsions to mortality
Data (dog/rat/mouse/hamster) from Weeks et al. 1963 450 ppm- hour 25% response (estimated) 300 ppm- hour 50% response 3,300 ppm- hour 66% response (estimated) Mode of toxic action: reactive metabolite causing uncharacterized central nervous system effects Allometric scaling: (body weight)$^{-0.75}$ Uncertainty factors: none Slope (rat): 14.7 Data plotted for humans: 450 ppm-hour 25% response 1,471 ppm-hour 50% response 4,950 ppm-hour 75% response Source: NRC 2000 *(Continued)*

1,1- and 1,2-Dimethylhydrazine (continued)
Mode of Action: Uncharacterized central nervous system effects
Dose-duration relationship: linear, $C^1 \times t = k$
Delayed sequellae: possible human carcinogen
For comparison: AEGL-1 = none AEGL-2 = 3 ppm-hour AEGL-3 = 11 ppm-hour

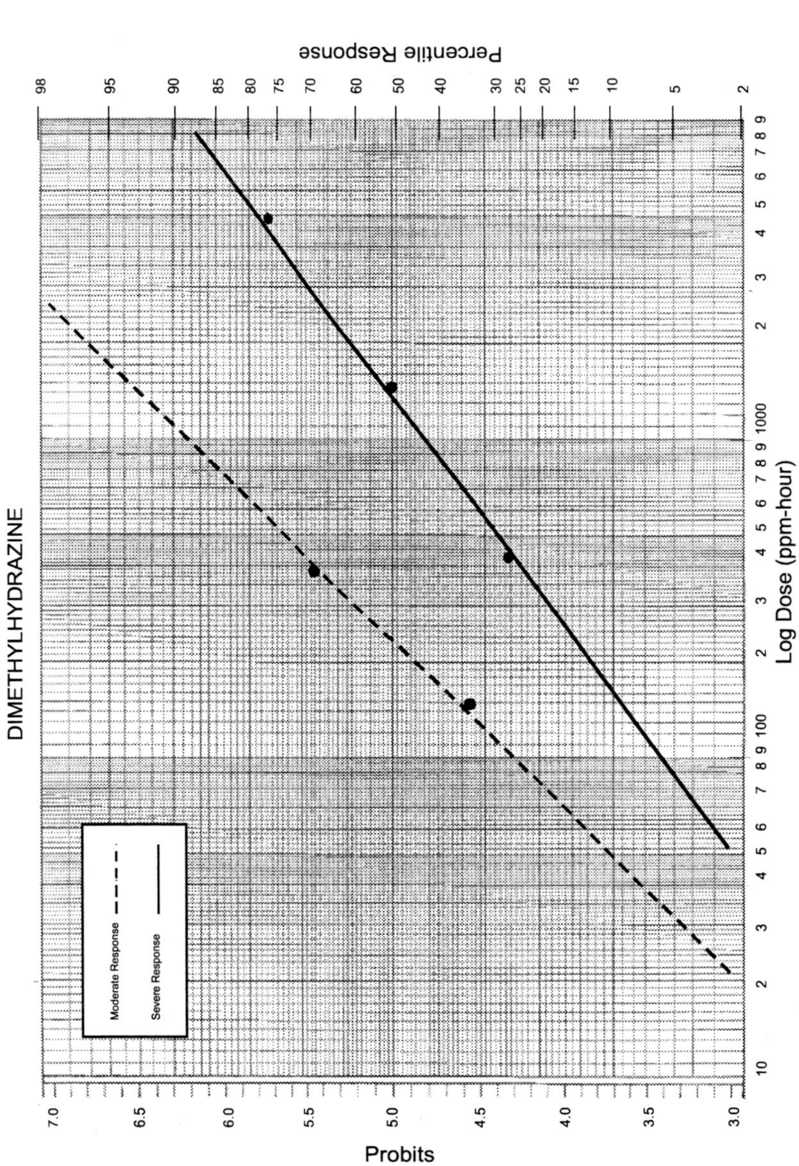
DIMETHYLHYDRAZINE
Log Dose (ppm-hour)
Probits
Percentile Response
Moderate Response
Severe Response

Hydrogen Cyanide
Mild Effects—headache, weakness
NOAEL (human) ≈ 5 ppm-8 hour/day, 5 day/week (NIOSH 1976) Data (human) from El Ghawabi 1975: 8 ppm-hour 25% response 10 ppm-hour 50% response 12 ppm-hour 75% response (estimated) Mode of toxic action: stable metabolite produces inhibition of cellular respiration Allometric scaling: not applicable Uncertainty factors: none Dose-duration relationship: log, $C^3 \times t = 1$ Data plotted for humans: 8 ppm-hour 25% response 10 ppm-hour 50% response 12 ppm-hour 75% response Source: NRC 2002
Moderate Effects—central nervous system depression
Data (monkey) from Purser 1984 30 ppm-hour 25% response (estimated) 35 ppm-hour 50% response (estimated) 40 ppm-hour 75% response (estimated) Mode of toxic action: stable metabolite produces inhibition of cellular respiration Allometric scaling: 1:1 for monkey:human Uncertainty factors: none Dose-duration relationship: log, $C^2 \times t = k$ (monkey) Data plotted for humans: 30 ppm-hour 25% response 35 ppm-hour 50% response 40 ppm-hour 75% response Source: NRC 2002 Time to incapacitation (monkey): 100 ppm 19 min 102 ppm 16 min 123 ppm 15 min 147 ppm 8 min 156 ppm 8 min *(Continued)*

Hydrogen Cyanide (continued)
Severe Effects—stimulation to depression to convulsions to coma to death
Data (rat) from E. I. duPont de Nemours Company 1981: 88 ppm- hour 10% response 108 ppm- hour 25% response (estimated) 139 ppm- hour 50% response 180 ppm-hour 75% response (estimated) Mode of toxic action: stable metabolite produces inhibition of cellular respiration Allometric scaling: 1:1 for rat:human Uncertainty factors: none Dose-duration relationship: log, $C^{2.6} \times t = k$ (rat); $C^2 \times t = k$ Data plotted for humans: 88 ppm-hour 10% response 108 ppm-hour 25% response 139 ppm-hour 50% response 180 ppm-hour 75% response Source: NRC 2002
Mode of Action: inhibition of cellular respiration
Dose-duration relationship: varies with level
Delayed sequellae: none known or anticipated
For comparison: AEGL-1 = 2.0 ppm-hour; 1.0 ppm-8 hour AEGL-2 = 7.1 ppm-hour; 2.5 ppm-8 hour AEGL-3 = 15 ppm-hour; 6.6 ppm-8 hour

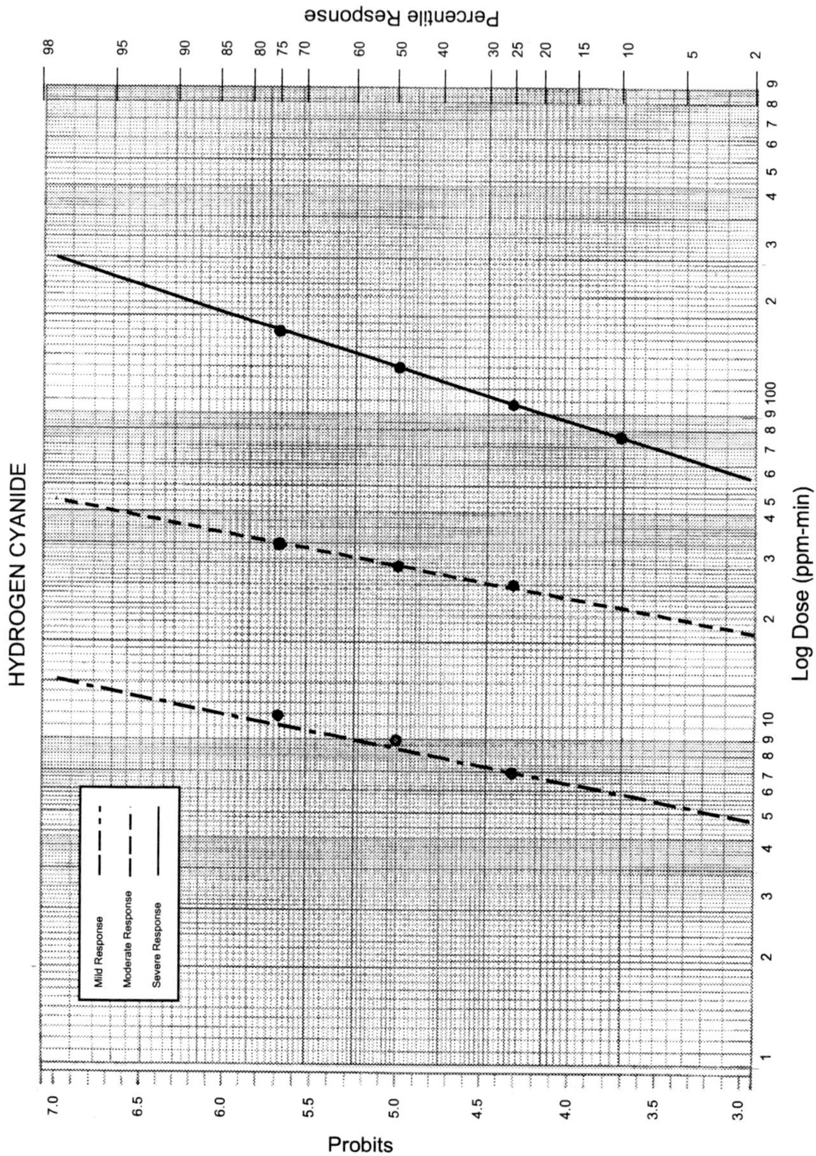
HYDROGEN CYANIDE
Percentile Response
Log Dose (ppm-min)
Probits
Mild Response
Moderate Response
Severe Response

Hydrogen Sulfide
Mild Effects—eye pain, photophobia, headache, irritation
Data (human) from WHO 1981 and Vanhoorne et al. 1995: 6 ppm-hour 0% response ? ppm-hour 50% response ? ppm-hour 75% response Mode of toxic action: direct effect on contact; edema systemically Allometric scaling: not applicable Uncertainty factors: none Dose-duration relationship: log, $C^{4.4} \times t = k$ (Note - may not apply to all asthmatic individuals) Data plotted for humans: 6 ppm-hour 0% response Source: EPA 2002
Moderate Effects: lacrymation, photophobia, corneal opacity, tracheobronchitis, central nervous system depression, nasal passage necrosis
Data: insufficient Dose-duration relationship: $C^n \times t = k$ (Note – might not apply to all asthmatic individuals) Source: EPA 2002
Severe Effects—cerebral and pulmonary edema to respiratory arrest to unconsciousness to death
Data (rat) from MacEwen and Vernot 1972: 635 ppm- hour 10% response 712 ppm- hour 50% response 800 ppm- hour 90% response Mode of toxic action: pulmonary and cerebral edema Allometric scaling: 1:1 for rat:human Uncertainty factors: none Data plotted for humans: 635 ppm-hour 10% response 712 ppm-hour 50% response 800 ppm-hour 90% response Source: EPA 2002
Mode of Action: inhibition of electron transport in tissues with high oxygen demand
Dose-duration relationship: varies with types of responses
Delayed sequellae: none known or anticipated
For comparison: AEGL-1 = 0.51 ppm-hour; 0.33 ppm-8 hour AEGL-2 = 27 ppm-hour; 17 ppm-8 hour AEGL-3 = 50 ppm-hour; 31 ppm-8 hour

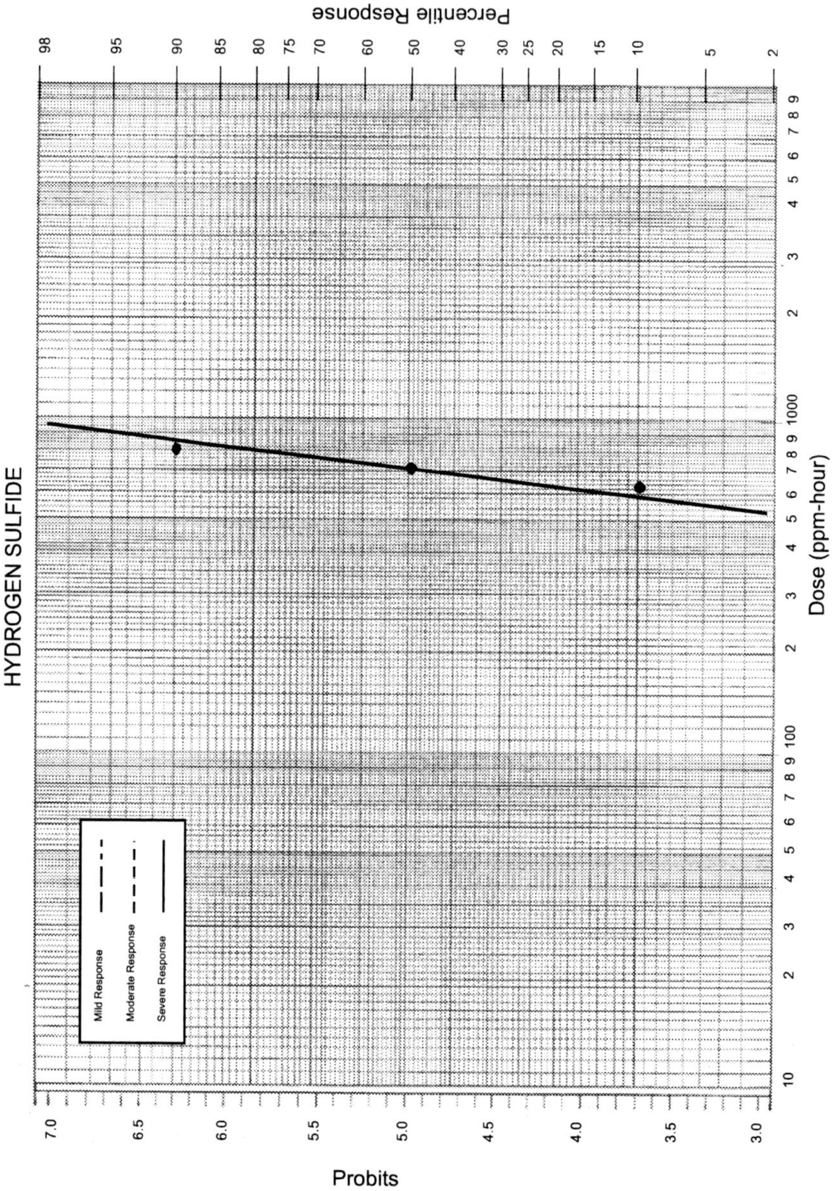
HYDROGEN SULFIDE
Percentile Response
98
95
90
85
80
75
70
60
50
40
30
25
20
15
10
5
2
Mild Response
Moderate Response
Severe Response
Dose (ppm-hour)
10
100
1000
Probits
7.0
6.5
6.0
5.5
5.0
4.5
4.0
3.5
3.0

Propylene Glycol Dinitrate
Mild Effects—headache
NOAEL (human) = 0.03 ppm for 8 hour Data (human) from Stewart et al. 1974: 0.1 ppm-hour 25% response 0.2 ppm-hour 50% response 0.4 ppm-hour 75% response Mode of toxic action: reactive metabolite produces vasodilation of cerebral vessels; decreased blood pressure Allometric scaling: not applicable Uncertainty factors: none Data plotted for humans: 0.1 ppm-hour 25% response 0.2 ppm-hour 50% response 0.4 ppm-hour 75% response Source: NRC 2002
Moderate Effects—severe headache, slight loss of equilibrium
Data (human) from Stewart et al. 1974 0.5 ppm-hour 25% response (estimated) 1.0 ppm-hour 50% response 2.0 ppm-hour 75% response (estimated) Mode of toxic action: reactive metabolite producing vasodilation of cerebral vessels; decrease in blood pressure Allometric scaling: not applicable Uncertainty factors: none Data plotted for humans: 0.5 ppm-hour 25% response 1.0 ppm-hour 50% response 2.0 ppm-hour 75% response Source: NRC 2002
Severe Effects—vomiting, central nervous system depression, semi-consciousness, clonic convulsions, mortality
Data (monkey) from Jones et al. 1972: 33 ppm-hour 25% response (estimated) 70 ppm-hour 50% response (estimated) 140 ppm-hour 75% response 280 ppm-hour 100% response (estimated) *(Continued)*

Propylene Glycol Dinitrate (continued)
Mode of toxic action: reactive metabolite producing decreased systolic pressure; increased diastolic pressure; myocardial eschimia; increased MetHb Allometric scaling: (body weight)$^{-0.75}$ Uncertainty factors: none Data plotted for humans: 60 ppm-hour 25% response 119 ppm-hour 50% response 238 ppm-hour 75% response Source: NRC 2002
Mode of Action: cardiovascular toxicity and central nervous system depression
Dose-duration relationship: linear, $C^1 \times t = k$; for severe, long-duration extrapolation $C^3 \times t = k$
Delayed sequellae: none identified
For comparison: AEGL-1 = 0.17 ppm-hour; 0.03 ppm-8 hour AEGL-2 = 1.0 ppm-hour; 0.13 ppm-8 hour AEGL-3 = 13 ppm-hour; 5.3 ppm-8 hour

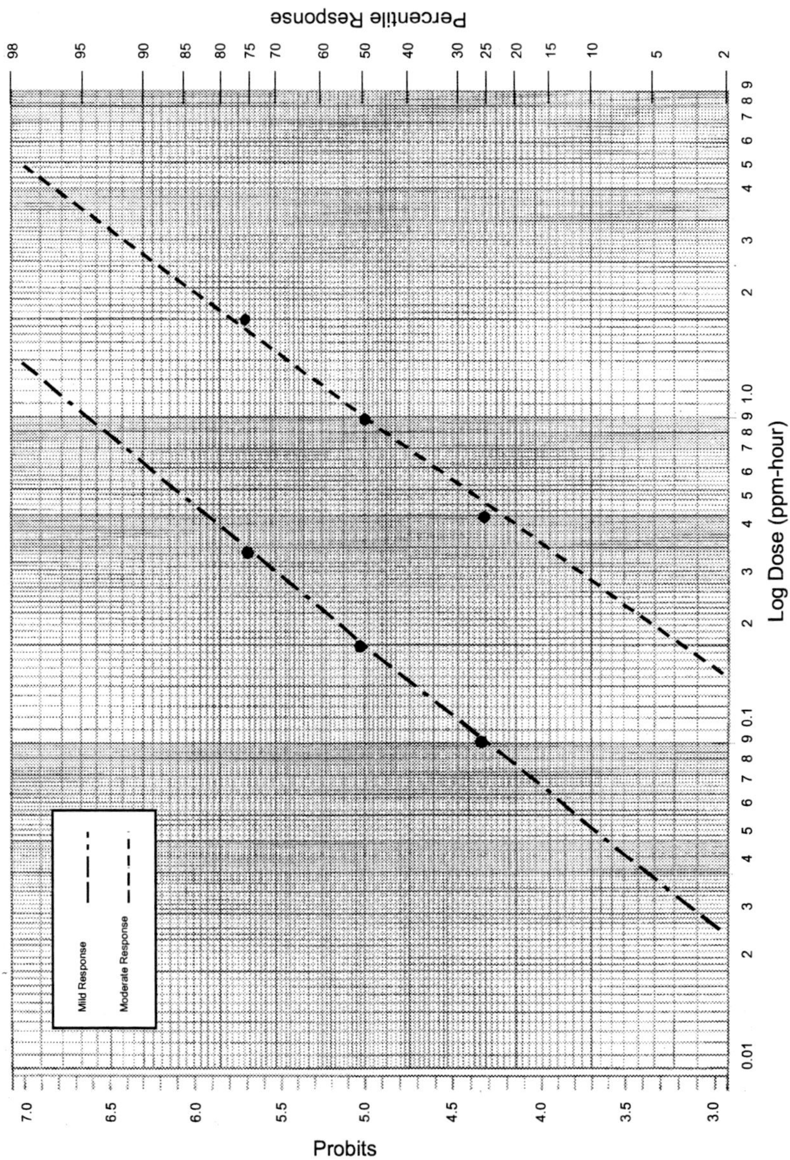

PROPYLENE GLYCOL DINITRATE
(PART 1)

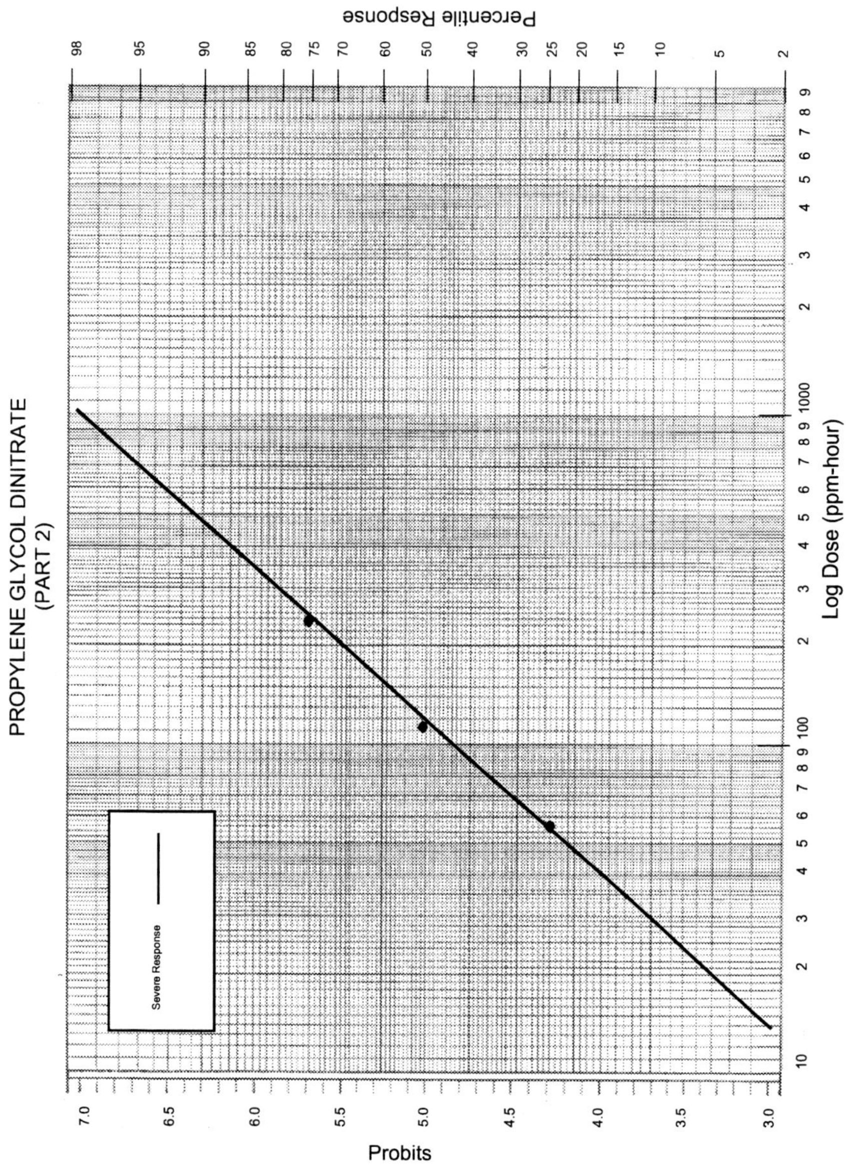
PROPYLENE GLYCOL DINITRATE
(PART 2)
Severe Response
Log Dose (ppm-hour)
Probits
Percentile Response

Sarin (GB)
Mild Effects—miosis, rhinorrhea
NOAEL (human) ≈ 0.016 mg/m^3 for 20 min Data (human) from NRC 2003: 0.32 mg-min/m^3 0% response ? mg-min/m^3 25% response 4 mg-min/m^3 50% response (ECT_{50}) ? mg-min/m^3 75% response Mode of toxic action: local Allometric scaling: not applicable Uncertainty factors: none Data plotted for humans: 0.32 mg-min/m^3 0% response 4 mg-min/m^3 50% response
Moderate Effects—insufficient data
Severe Effects—acetylcholinesterase inhibition to convulsions to mortality
Data (monkey) from NRC 2003: 1 mg-hour/m^3 1% response (estimated from rat data) ? mg-hour/m^3 25% response 27-150 mg-hour/m^3 50% response ? mg-min/m^3 75% response Mode of toxic action: stable metabolite leading to acetylcholinesterase inhibition Allometric scaling: 1:1 for monkey:human Uncertainty factors: none Data plotted for humans: 1 mg-hour/m^3 1% response 90 mg-hour/m^3 50% response
Mode of Action: inhibition of acetylcholinesterase leading to convulsions and then to death
Dose-duration relationship: linear, $C^2 \times t = k$ (?)
Delayed sequellae: delayed neuropathy
For comparison: AEGL-1 = 0.0028 mg-hour/m^3 AEGL-2 = 0.035 mg-hour/m^3 AEGL-3 = 0.13 mg-hour/m^3

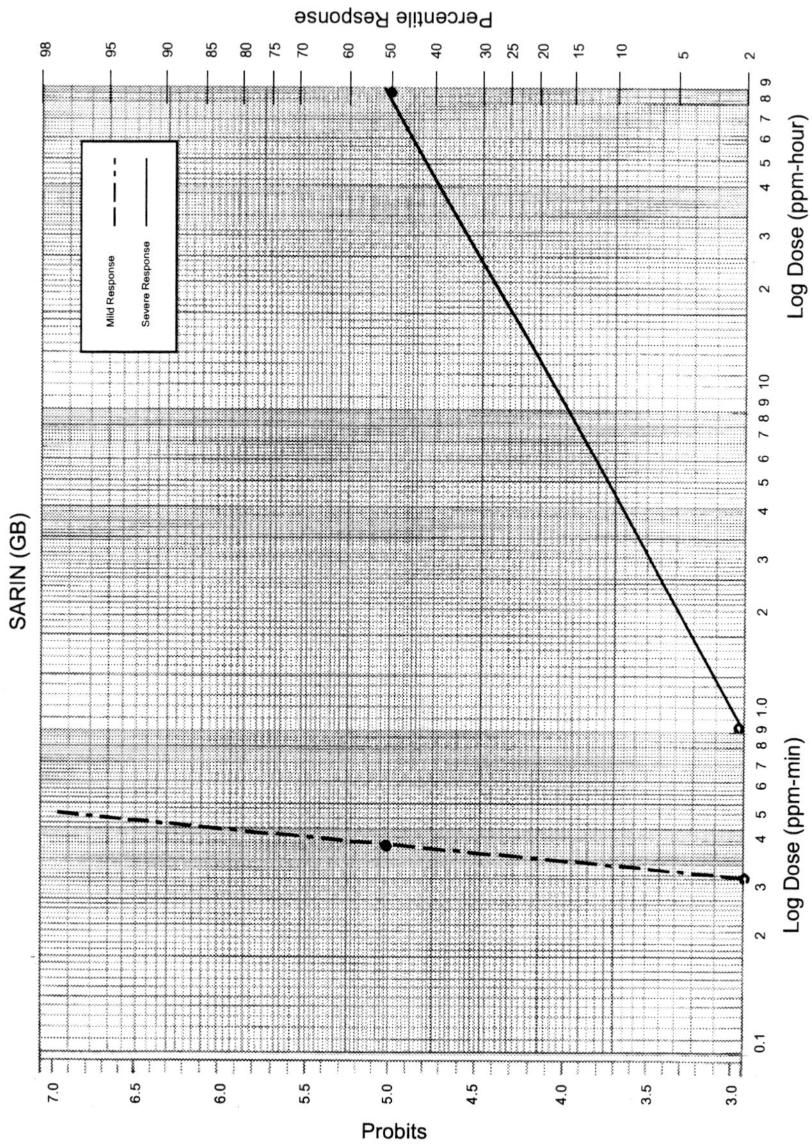
SARIN (GB)
Mild Response
Severe Response
Percentile Response
Probits
Log Dose (ppm-min)
Log Dose (ppm-hour)

REFERENCES

Bodansky, O. 1951. Methemoglobinemia and methemoglobin-producing compounds. Pharmacol. Rev. 3:144-196.

Clewell, H., III., M.E. Andersen, and H.A. Barton. 2002. A consistent approach for the application of pharmacokinetic modeling in cancer and noncancer risk assessment. Environ. Health Perspect. 110(1):85-93.

E.I. duPont de Nemours Company. 1981. Inhalation Toxicity of Common Combustion Gases. Report No. 238-81. Haskell Laboratory, Newark, DE.

El Ghawabi, S.H., M.A. Gaafar, A.A. El-Saharti, S.H. Ahmed, K.K. Malash, and R. Fares. 1975. Chronic cyanide exposure: A clinical, radioisotope, and laboratory study. Br. J. Ind. Med. 32(3):215-219.

EPA (U.S. Environmental Protection Agency). 1994. Methods for Derivation of Inhalation Reference Concentrations and Application of Inhalation Dosimetry. EPA/600/8-90/066F. Environmental Criteria and Assessment Office, Office of Health and Environmental Assessment, Office of Research and Development, .S. Environmental Protection Agency, Research Triangle Park, NC. October 1994.

EPA (U.S. Environmental Protection Agency). 2002. Hydrogen Sulfide, Interim Acute Exposure Guideline Levels (AEGLs). Interim 4 Technical Support Document: 11/2002. Office of Pollution Prevention and Toxics, U.S. Environmental Protection Agency.

Jones, R.A., J.A. Strickland, and J. Siegel. 1972. Toxicity of propylene glycol 1,2-dinitrate in experimental animals. Toxicol. Appl. Pharmacol. 22(1):128-137.

Kiese, M. 1974. Methemoglobinemia: A Comprehensive Treatise; Causes, Consequences, and Correction of Increased Contents of Ferrihemoglobin in Blood. Cleveland, OH: CBC Press.

Kim, Y.C., and G.P. Carlson. 1986. The effect of an unusual workshift on chemical toxicity. II. Studies on the exposure of rats to aniline. Fundam. Appl. Toxicol. 7(1):144-152.

MacEwen, J.D., and E.H. Vernot. 1972. Pp. 66-69 in Toxic Hazards Research Unit Annual Report: 1972. Report No. ARML-TR-72-62. NTIS AD755-358. Aerospace Medical Research Laboratory, Air Force Systems Command, Wright-Patterson Air Force Base, OH.

NIOSH (National Institute for Occupational Safety and Health). 1976. Occupational Exposure to Hydrogen Cyanide and Cyanide Salts (NaCN, KCN, and $Ca(CN)_2$): Criteria for a Recommended Standard. DHEW (NIOSH) 77-108. Cincinnati: U.S. Department of Health, Education, and Welfare, Public Health Service, Center for Disease Control, National Institute for Occupational Safety and Health.

NRC (National Research Council). 2000. Acute Exposure Guideline Levels for Selected Airborne Chemicals, Vol. 1. Washington, DC: National Academy Press.

NRC (National Research Council). 2001. Standard Operating Procedures for Developing Acute Exposure Guideline Levels for Hazardous Chemicals. Washington, DC: National Academy Press.

NRC (National Research Council). 2002. Acute Exposure Guideline Levels for Selected Airborne Chemicals, Vol. 2. Washington, DC: The National Academies Press.

NRC (National Research Council). 2003. Acute Exposure Guideline Levels for Selected Airborne Chemicals, Vol. 3. Washington, DC: The National Academies Press.

Purser, D.A. 1984. A bioassay model for testing the incapacitating effects of exposure to combustion product atmospheres using cynomolgus monkeys. J. Fire Sci. 2:20-36.

Seger, D.L. 1992. Methemoglobin-forming chemicals. Pp. 800-806 in Hazardous Materials Toxicology: Clinical Principles of Environmental Health, J.B. Sullivan and G.R. Krieger, eds. Baltimore, MD: Williams & Wilkins.

Sim, V.M., and R.E. Pattle. 1957. Effect of possible smog irritants on human subjects. J. Am. Med. Assoc. 165(15):1908-1913.

Smyth, H.F., Jr. 1956. Hygienic standards for daily inhalation. Am. Ind. Hyg. Assoc. Q 17(2):129-185.

Stephens, E.R., E.F. Darley, O.C. Taylor, and W.E. Scott. 1961. Photochemical reaction products in air pollution. J. Air Pollut. 4:79-100.

Stewart, R.D., J.E. Peterson, P.E. Newton, C.L. Hake, M.J. Hosko, A.J. Lebrun, and G.M. Lawton, 1974. Experimental human exposure to propylene glycol dinitrate. Toxicol. Appl. Pharmacol. 30(3):377-395.

USACHPPM (U.S. Army Center for Health Promotion and Preventive Medicine). 2002a. Chemical Exposure Guidelines for Deployed Military Personnel. Technical Guide 230. U.S. Army Center for Health Promotion and Preventive Medicine. January 2002. [Online]. Available: http://chppm-www.apgea.army.mil/deployment/ [accessed November 25, 2003]

USACHPPM (U.S. Army Center for Health Promotion and Preventive Medicine). 2002b. Chemical Exposure Guidelines for Deployed Military Personnel. A Companion Document to USACHPPM Technical Guide (TG) 230 Chemical Exposure Guidelines for Deployed Military Personnel. Reference Document (RD) 230. U.S. Army Center for Health Promotion and Preventive Medicine January 2002. [Online]. Available: http://chppm-www.apgea.army. mil/ deployment/ [accessed November 25, 2003]

Vanhoorne, M., A. de Rouck, and D. de Bacquer. 1995. Epidemiological study of eye irritation by hydrogen sulfide and/or carbon disulphide exposure in viscose rayon workers. Ann. Occup. Hyg. 39(3):307-315.

Weeks, M.H., G.C. Maxey, M.E. Sicks, and E.A. Greene. 1963. Vapor toxicity of UDMH in rats and dogs from short exposures. Am. Ind. Hyg. Assoc. J. 24:137-143.

WHO (World Health Organization). 1981. Hydrogen Sulfide. Environmental Health Criteria 19. Geneva: World Health Organization.

Appendix D

Critical Studies and Uncertainty Factors Used in Developing Acute Exposure Guideline Levels for Chemical Warfare Agents

TABLE D-1 Critical Studies and Uncertainty Factors Used in Developing AEGLs for GB

Critical Study							Uncertainty Factors (UFs)			
Standard and Study	End Point	Severity	Species	Gender	Test Duration	Temporal Extrapolation	Intra-species	Inter-species	MF	Total UF
AEGL-1 Mioduszewski et al. 2000	EC_{50} for miosis	Nondisabling (miosis is the first measurable change in the continuum of response to anticholinesterase response)	Adult rat	Female	10, 60, and 240 min	$C^n \times t = k$; n = 2; from 10 to 30 min and from 4 to 8 hours	10	1	1	10
AEGL-2 Baker and Sedgwick 1996	LOAEL for miosis, dyspnea, RBC-ChE inhibition, single fiber electromyography changes	Disabling (single fiber electro-myography changes as early indicator of exposures that could result in more significant effects)	Human	NR	10-30 min	$C^n \times t = k$; n = 2; from 30 min to all durations	10	1	1	10
AEGL-3 Mioduszewski et al. 2000, 2001, 2002	LC_{01} and LC_{50}	Lethal	Rat	Female	10, 30, 60, 90, 240, and 360 min	$C^n \times t = k$; n = 2; from 6 to 8 hours	10	3	1	30

Abbreviations: EC_{50}, effective concentration for a 50% response; LC_{01}, lethal concentration to 1% of exposed subjects; LC_{50}, lethal concentration to 50% of exposed subjects; MF, modifying factor; NR, not reported; RBC-ChE, red blood celll cholinesterase.
Source: Data obtained from NRC 2003.

TABLE D-2 Relative Potency and Uncertainty Factors used in Developing AEGLs for GA, GD, GF, and VX

			Uncertainty Factors (UFs)			
Agent	Standard	Relative Potency[a]	Intraspecies	Interspecies	MF	Total
GA	AEGL-1 and AEGL-2	Equivalent potency	10	1	1	10
	AEGL-3	GA = ½ × GB potency	10	3	1	30
GD	AEGL-1 and AEGL-2	GD = 2 × GB potency	10	1	1	10
	AEGL-3	Equipotent to GB; supported by Wistar rat LC_{50} study	10	3	1	30
GF	AEGL-1 and AEGL-2	GF = 2 × GB potency	10	1	1	10
	AEGL-3	Equipotent to GB	10	3	1	30
VX	AEGL-1	VX:GB = 4; miosis data from secondary and supportive studies	10	1	3	30
	AEGL-2 and AEGL-3	VX:GB = 4	10	3	3	100

[a]Based on relative potency data from GB.
Abbreviations: LC_{50}, lethal concentration to 50% of subjects exposed; MF, modifying factor.
Source: Data obtained from NRC 2003.

TABLE D-3 Critical Studies and Uncertainty Factors Used in Developing AEGLs for Sulfur Mustard

Critical Study						Uncertainty Factors (UFs)			
Standard and Study	End Point	Severity	Species	Test Duration	Temporal Extrapolation	Intra-species	Inter-species	MF	Total UF
AEGL-1 Anderson 1942	Threshold—conjunctival injection and minor discomfort with no functional decrement	Nondisabling	Human		$C^n \times t = k$; $n = 1$	3	1	1	3
AEGL-2 Anderson 1942	LOAEL for well-marked, generalized conjunctivitis, edema, photophobia, and eye irritations	Severe ocular effects; ineffective military performance	Human		$C^n \times t = k$; $n = 1$	3	1	3[a]	10
AEGL-3 Kumar and Vijayaraghavan 1998	LC_{50}	Lethality	Mice	1 hour	$C^n \times t = k$; $n = 3$ for shorter periods and $n = 1$ for longer periods	3	3	1	10

[a]Potenial onset of long-term ocular or respiratory effects.
Abbreviations: LC_{50}, lethal concentration to 50% of exposed subjects; LOAEL, lowest-observed-adverse-effect level; MF, modifying factor.
Source: Data obtained from NRC 2003.

REFERENCES

Anderson, J.S. 1942. The Effect of Mustard Gas Vapour on Eyes under Indian Hot Weather Conditions. CDRE Report No. 241. Chemical Defense Research Establishment (India).

Baker, D.J., and E.M. Sedgwick. 1996. Single fibre electromyographic changes in man after organophosphate exposure. Hum. Exp. Toxicol. 15(5):369-375.

Kumar, O., and R. Vijayaraghavan. 1998. Effect of sulphur mustard inhalation exposure on some urinary variables in mice. J. Appl. Toxicol. 18(4):257-259.

Mioduszewski, R.J., J. Manthei, R. Way, D. Burnett, B. Gaviola, W. Muse, S. Thomson, D. Sommerville, and R. Crosier. 2000. Estimating the probability of sarin vapor toxicity in rats as a function of exposure concentration and duration. Proceedings of the International Chemical Weapons Demilitarization Conference (CWD-2000), May 21-24, 2000, The Hague, NL.

Mioduszewski, R.J., J. Manthei, R. Way, D. Burnett, B. Gaviola, W. Muse, J. Anthony, D. Durst, D. Sommerville, R. Crosier, S. Thomson, and C. Crouse. 2001. ECBC Low Level Operational Toxicology Program: Phase I-Inhalation Toxicity of Sarin Vapor in Rats as a Function of Exposure Concentration and Duration. ECBC-TR-183. Edgewood Research Development and Engineering Center, Aberdeen Proving Ground, MD. (August 2001).

Mioduszewski, R.J., J. Manthei, R. Way, D. Burnett, B. Gaviola, W. Muse, S. Thomson, D. Sommerville, R. Crosier, J. Scotto, D. McCaskey, C. Crouse, and K. Matson. 2002. Low-Level Sarin Vapor Exposure in Rats: Effect of Exposure Concentration and Duration on Pupil Size. ECBC-TR-235. Edgewood Chemical Biological Center, U.S. Army Soldier and Biological Chemical Command Aberdeen Proving Ground, MD. (May 2002).

NRC (National Research Council). 2003. Acute Exposure Guideline Levels for Selected Airborne Chemicals, Volume 3. Washington, DC: The National Academies Press.

Appendix E

Probabilistic Approach to Address Exposure to Multiple Chemicals for Course-of-Action Analysis

The percent P^* of a unit expected to be seriously affected (in a mission-hindering way) by acute respiratory exposure(s) to multiple chemicals could be estimated using the following procedure. Such exposures might affect similar toxic end points and/or different toxic end points. This procedure is a generalization of the assumption made in Chapter 4 that, in the case of exposure to a single chemical, $P^* = F \times P$ where F is the estimated fraction of the unit exposed to that chemical and P is an estimate of the percent of exposed individuals expected to incur serious (mission-incapacitating) illness, modeled as

$$P = \Phi\left[1.036\frac{\log(C / C_{50})}{\log(C_{50} / C_{15})}\right] \times 100\%,$$
$$P = \Phi\left[\frac{\log(C / C_{50})}{\sigma}\right] \times 100\%. \qquad \text{(E-1)}$$

In that equation, Φ is the cumulative normal (Gaussian) probability distribution function; log denotes logarithm (using any specified base, such as 10 or e); C_{50} and C_{15} are estimated (e.g., 1-hour) concentrations that elicit a 50% and 15% response, respectively, obtained from a lognormal dose-response curve previously fitted to relevant toxicity data as described in Appendix C (estimates of C_{50} and C_{15} are listed for five chemicals in Chapter

4, Table 4-1); and σ is the estimate of the lognormal-model "shape" parameter, the inverse of which specifies the steepness of the corresponding fitted dose-response curve.

The subcommittee wishes to consider the general case in which deployed personnel are exposed to n subsets of chemicals, each ith subset of which contains n_i chemicals (for $i = 1, \ldots, n$) that all induce a common (mechanism-specific) toxic end point T_i that is independent from any different T_i–specific mechanisms and from any different end points T_j (for $j \neq i$) induced by other chemicals involved in the exposure scenario considered. Typically neither n nor n_i will be large (i.e., exceed 2 or 3), but the case of multiple chemicals is treated here in general terms to explain the general approach clearly. The 1-hour respiratory concentration of the jth chemical in the ith subset shall be denoted $C_{i,j}$, where $j = 1, \ldots, n_i$ and where again $i = 1, \ldots, n$. Because this general treatment involves double-subscript notation, it will be convenient to adopt the alternative notation $\mu = C_{50}$ for the lognormal location-parameter estimate C_{50}. Thus, for the jth chemical in the ith subset, Equation E-1 can be rewritten as

$$P_{i,j} = \Phi\left[\frac{\log(C_{i,j} / \mu_{i,j})}{\sigma_{i,j}}\right] \times 100\%. \qquad \text{(E-2)}$$

Different chemicals with similar values of σ as well as similar values of μ for a given end point can be treated as if they were all the same chemical. In the absence of information supporting an alternative assumption, it is reasonable to assume in the context of using the lognormal dose-response model that different chemicals with substantially different values of σ for a given end point act via independent mechanisms. Concentrations $C_{i,j}$ of different chemicals affecting a common end point T_i, all of which have similar values of $\sigma_{i,j}$ but have substantially different values of $\mu_{i,j}$, can reasonably be assumed to represent corresponding weighted contributions to an aggregate effective concentration C_i that acts via a single underlying mechanism to elicit T_i. We now assume that, for all i, the ith set of median-response concentrations $\mu_{i,j}$ are sorted in order of ascending magnitude for $j = 1, \ldots, n_i$, such that $\mu_{i,1}$ always denotes the smallest median-response concentration, which in turn corresponds to that chemical with the highest toxic potency among the ith subset of chemicals. It follows that the relative weight of the jth contribution to C_i is $(\mu_{i,1}/\mu_{i,j})$, and thus that

$$C_i = \sum_{j=1}^{n_i} C_{i,j} (\mu_{i,1} / \mu_{i,j})$$

The incapacitated percent P_i of a unit exposed to the *i*th subset of chemicals is therefore approximated by

$$P_i \approx \Phi\left[\frac{\log(C_i / \mu_{i,1})}{\mathrm{M}(\sigma_{i,j})}\right] \times 100\% = \Phi\left[\frac{\log\left(\sum_{j=1}^{n_i} C_{i,j} / \mu_{i,j}\right)}{\mathrm{M}(\sigma_{i,j})}\right] \times 100\%, \quad \text{(E-3)}$$

where the function M in the denominator of each bracketed expression denotes the arithmetic mean over $j = 1, \ldots, n_j$ with similar values of $\sigma_{i,j}$ conditional on i. A more conservative estimate of P_i is obtained if M is taken to be the maximum or the minimum of $\sigma_{i,j}$ when the numerator of each bracketed expression in Equation E-3 is ≤ 0 or is >0, respectively.

The probability P of incapacitation via *any* of the n end points considered is (by "de Morgan's rule") equal to the complement of the joint probability that *none* of these n end points will occur. Because the probability of incapacitation via (mechanism-specific) end point T_i is by definition independent of that via end point T_j for $i \neq j$, it follows that this joint probability is just the product of the probabilities $(1 - P_i)$ for $i = 1, \ldots, n$, and consequently that

$$P = 100\% - \prod_{i=1}^{n} (100\% - P_i), \quad \text{(E-4)}$$

where P_i was defined in approximation E-3.

The percentage P^* of a unit seriously affected by chemical exposure is finally calculated as $P^* = P \times F$, where F is the estimated fraction of the unit exposed to the n subsets of chemicals considered. If each among m mutually exclusive fractions F_k of a unit (for $k = 1, \ldots, m$) is exposed to a combination of chemical concentrations that together generate a corresponding predicted response percentage $P_{[k]}$ calculated using Equation E-4 (where bracket-subscript notation is used to distinguish this percentage from one defined by Equation E-3), then

$$P^* = \sum_{k=1}^{m} F_k P_{[k]}.$$

As explained in Chapter 4, the calculated percentage P^* directly specifies the unit status as indicated in Chapter 2 (Table 2-4). In this way, P^* is used to classify the operational risk management (ORM) risk level defined in the military risk assessment matrix by comparing the quantity (100% – P^*) directly to the unit-strength percentage ranges that define the various risk levels as specified in Appendix C of FM 101-5-1, in TG 230 Table 3-4, and in Chapter 2 (Table 2-4) and Chapter 4 (Table 4-1) of this report.

This default probabilistic risk-modeling approach for multiple chemicals explained above is illustrated in Box E-1.

BOX E-1 Example of Probabilistic Approach for Multiple Chemicals

It is assumed that a proposed mission would require 40% of a unit be exposed by inhalation for 1 hour to simultaneous ambient concentrations of 25, 4, and 2 ppm of hydrogen sulfide, hydrogen cyanide, and 1,1-dimethylhydrazine, respectively. From Chapter 4 (Table 4-1), corresponding information listed in Appendix C, and the definitions $\mu = C_{50}$ and $\sigma = 0.9648 \log (C_{50}/C_{15})$, the exposure levels, the lognormal-model parameter estimates μ and σ, and the corresponding "severe" toxic end points for these three chemicals can be summarized as follows:

Chemical	Subset i	j	$C_{i,j}$ (ppm-hour)	$\mu_{i,j}$ (ppm-hour)	$\sigma_{i,j}$ (unitless)	End Point T_i
Hydrogen cyanide	1	1	60	140	0.37	Hypoxia
Hydrogen sulfide	1	2	300	710	0.1	Hypoxia
Dimethylhydrazine	2	1	450	1400	1.66	CNS

Severe effects of respiratory exposure to either hydrogen cyanide or hydrogen sulfide include severe and potentially lethal histotoxic hypoxia due to inhibition of cellular oxidative metabolism, and σ? values for these chemicals are both relatively small. In contrast, severe effects of dimethylhydrazine exposure include tremors and vomiting via uncharacterized central nervous system (CNS) interference. This example could be considered to involve $n = 2$ subsets of chemicals: the first including hydrogen cyanide and hydrogen sulfide (the two chemicals with a similar toxicity mechanism, with σ? ≈ 0.374), and the second including only dimethylhydrazine. Corresponding application of Equation C-2, Equation C-3, and the definition $P^* = P \times F$ with $F = 0.40$, in this case

$$P_1 \approx \Phi\left[0.347^{-1} \log_e\left(\frac{60}{140} + \frac{300}{710}\right)\right] \times 100\% \approx 33.3\%,$$

$$P_2 = \Phi\left[1.66^{-1} \log_e\left(\frac{450}{1{,}400}\right)\right] \times 100\% \approx 24.7\%,$$

$$P = 100\% - (100\% - 33.3\%)(100\% - 24.7\%) \approx 49.8\%,$$

$$P^* = 0.40 \times P \approx 20\%.$$

(Continued)

BOX E-1 continued

yields which implies that the proposed mission has a unit status of "amber" (mission capable, with minor deficiencies implying a total of 70-85% unit strength) if chemical exposures were considered the only risks to the mission. If in this example toxicity due to hydrogen cyanide and hydrogen sulfide were considered to arise from completely independent mechanisms, then corresponding calculations would yield $P_1 \approx 1.2\%$, $P_2 \approx 0\%$ and $P_3 \approx 24.7\%$, implying that $P \approx 25.6\%$ and $P^* \approx 10\%$ and consequently, that the mission has a unit status of green (mission capable, with unit strength $\geq 85\%$). An additive approximation of aggregate risk due to both hydrogen cyanide and hydrogen sulfide assuming complete independence would be 1.2% + ~0% ≈ 1.2%. Note how much the latter approach underestimates the corresponding aggregate risk (~33%) that was predicted above assuming a common mechanism of action and a (conservatively estimated) common value of σ.

Abbreviations: AEGL, acute exposure guideline level; CNS, central nervous; ppm, parts per million.

Appendix F

Biographical Information on the Subcommittee on Toxicological Risks to Deployed Military Personnel

RICHARD J. BULL (*Chair*) is professor of environmental science at Washington State University TriCities. His research interests include the toxicology of drinking water disinfection byproducts and halogenated solvents. He has been involved in health-risk assessments of hazardous waste sites and other chemical hazards. He is also part of a major effort to integrate new findings in reduction biology into a more comprehensive approach to cancer risk assessment. Dr. Bull worked for 14 years at the U.S. Environmental Protection Agency's (EPA's) Health Effects Research Laboratory, where he held a number of positions, including director of the toxicology and microbiology division, and he is a former senior staff scientist with Battelle Pacific Northwest Laboratory. He has served as an advisor on many national scientific advisory committees, including service as chair of EPA's Drinking Water Committee and chair of the National Research Council's (NRC's) Committee on Copper in Drinking Water. Dr. Bull received his Ph.D. in pharmacology from the University of California, San Francisco.

EDWARD BISHOP is vice president of Parsons Corporation. He served as an officer in the U.S. Air Force and has 26 years of experience as an industrial hygienist and environmental engineer. His work experience is in the areas of environmental compliance, remedial investigations, hazardous waste minimization, industrial process evaluation, pollution prevention, industrial hygiene, and risk assessment. Dr. Bishop is a member of the NRC Committee on Toxicology and the Subcommittee on Acute Exposure Guideline Levels. He received his M.S. in engineering from the University of California, Los Angeles, and his Ph.D. in environmental health sciences from the University of California, Berkeley.

KENNETH T. BOGEN is a senior environmental scientist at Lawrence Livermore National Laboratory's Environmental Science Division at the University of California, Livermore. His research involves cancer-risk assessment methods, regulatory toxicology, biodosimetric and pharmacokinetic modeling, and quantitative uncertainty analysis. He has been a principal and co-investigator on related research projects funded by the U.S. Department of Energy, EPA, and others. He is an appointed member of the University of California Davis Cancer Center, and past president and current councilor of the Northern California Chapter of the Society for Risk Analysis. He served in 2000-2001 as chairman of the U.S. Consumer Product Safety Commission's Chronic Hazard Advisory Panel (CHAP) on Diisononyl Phthalate (DINP). Dr. Bogen served on the NRC Committee on Science and Judgment in Risk Assessment. He received his Dr.P.H. at the University of California, Berkeley.

BARBARA G. CALLAHAN is a senior toxicologist at University Research Engineers and Associates, and also holds an appointment as adjunct associate professor in environmental health sciences at the University of Massachusetts, Amherst. Her research interests include exposure and risk assessment evaluations of sites contaminated with pesticides, PCBs, heavy metals, and polycyclic aromatic hydrocarbons. She is a member of several national committees that study the effects of acute exposure to toxicants on human health after accidental release under emergency conditions and is senior editor for *Human and Ecological Risk Assessment*. Dr. Callahan is a recipient of the U.S. Army Environmental Hygiene Agency Commander's Medallion. She is former member of the NRC's Standing Committee on Program and Technical Review of the U.S. Army Chemical and Biological Defense Command. Dr. Callahan received her M.S. in biology from Rivier College and her Ph.D. in toxicology from Northeastern University. She is also a diplomate of the American Board of Toxicology.

JUDITH GRAHAM is a senior scientist with the American Chemistry Council (ACC). She serves as senior director of the council's long-range research initiative (LRI) team that sponsors research that advances the science of risk assessment for the health and ecological effects of chemicals to support decision making by government, industry, and the public. Her research interests include inhalation toxicology, exposure analysis, and health effects and health risks of air pollutants. Before joining ACC, Dr. Graham was with EPA for 32 years. Her last position was associate director for health at EPA's National Exposure Research Laboratory (NERL). She is a former president of the Inhalation Specialty Section and the Risk Assessment Specialty Section of the Society of Toxicology; the International Society of Exposure Analysis; and the Academy of Toxicological Sciences. She is a member of the NRC Committee on Toxicology. Dr. Graham received her Ph.D. in physiology and pharmacology from Duke University.

DAVID H. MOORE is vice president of defense medical technology at Battelle Eastern Science and Technology Center. Before joining Battelle, he served for over 20 years as a scientist in U.S. Army medical research and development. He retired as deputy director of the U.S. Army Medical Research Institute of Chemical Defense. Dr. Moore was involved in elucidating the effects of nerve agents on airway smooth muscle, developed the concept of a topical skin protectant, and published a number of papers on the pharmacokinetics of oximes and anticonvulsants for treated nerve-agent poisoning. He served on the Institute of Medicine's (IOM's) Committee on Research and Development Needs for Improved Civilian Medical Response to Chemical and Biological Terrorism Incidents and the NRC's Deployed Forces Advisory Group. Dr. Moore is currently a member of the NRC Committee on Toxicology. He is also currently serving on panels for the Naval Studies Board and the Air Force Science Advisory Board. Dr. Moore received his D.V.M. from the University of Georgia and his Ph.D. in physiology from Emory University.

DEBORAH IMEL NELSON is associate professor in the School of Civil Engineering and Environmental Science at the University of Oklahoma. Her research interests include occupational and environmental health risk assessment and the development of risk-based occupational exposure limits. She recently served for 2 years as an occupational health scientist with the World Health Organization, where she coordinated the Global Burden of Occupational Disease and Injury Project and conducted all of the exposure assessments for the project. Dr. Nelson has held a number of leadership positions in the American Industrial Hygiene Association, including service on the board of directors, co-founder and former chair of the association's risk-assessment committee, and secretary of the board. Dr. Nelson received her M.E.S. in environmental science and her M.P.H. and Ph.D. in environmental health from the University of Oklahoma. She is a certified industrial hygienist.

CHARLES F. REINHARDT retired in 1996 from DuPont's Haskell Laboratory, where he spent 30 years in a number of positions, including the directorship of the laboratory from 1976 to 1996. He is past president of the American Board of Toxicology and the American College of Occupational and Environmental Medicine. He is certified by the American Board of Preventive Medicine in occupational medicine and by the American Board of Toxicology in general toxicology. Dr. Reinhardt currently serves on the NRC Committee on the Review and Evaluation of the Army Chemical Stockpile Disposal Program. He received his M.D. from Indiana University's School of Medicine and M.Sc. in occupational medicine from Ohio State University.

ROSALIND A. SCHOOF is a consultant in toxicology and risk assessment with Integral Consulting, Inc. Dr. Schoof has extensive toxicology consulting experience and previously worked for a pharmaceutical company, where she developed safety assessment research programs for new drug candidates. Dr. Schoof has conducted evaluations of environmental chemical toxicity, health-risk assessments for cancer and noncancer end points, and multimedia assessments of exposure to environmental chemicals at diverse manufacturing sites, including brownfield sites and military installations. Dr. Schoof's particular research interests include the bioavailability of metals present in soils and dietary exposures to metals. She was a member of the NRC Committee on Toxicants and Pathogens in Biosolids Applied to Land. She received her Ph.D. in toxicology from the University of Cincinnati and is a diplomate of the American Board of Toxicology.

ROBERT TARDIFF is president of The Sapphire Group, Inc., a consulting group that focuses on hazard assessment, chemical interactions, risk assessment, risk communication, and risk management. He has held a number of senior positions in other consulting organizations, including EA Engineering, Science and Technology, Versar, Inc., and Environ Corporation. He was also chief of the Toxicological Assessment Branch of EPA between 1970 and 1977. Dr. Tardiff is a former president of the Society for Risk Analysis and is an editor of several toxicology and environmental health journals. He received his Ph.D. in toxicology and pharmacology from the University of Chicago.

NGA TRAN is a senior managing scientist at Exponent, Inc., and is an adjunct assistant professor at the Johns Hopkins University Bloomberg School of Public Health. She formerly held the position of special assistant to the assistant secretary of the Office of Environmental Safety and Health at the U.S. Department of Energy. Her research interests include health-risk assessment, risk management, and risk-based priority setting. Dr. Tran received her M.P.H. from Yale University, M.B.A. from DePaul University, and Dr.P.H. from Johns Hopkins University. She is also a certified industrial hygienist with chemical industry experience.

Appendix G

Definitions

Aggregate exposure. Exposure to a single chemical by multiple pathways (e.g., air, food, drinking water) and routes of exposure (inhalation, oral, and dermal).

Benchmark dose (BMD). Dose with a specified low level of excess health risk, generally in the range of 1% to 10%, which can be estimated from data with little or no extrapolation outside the experimental dose range. It is derived by modeling the data in the observed experimental range, selecting an incidence level within or near the observed range (e.g., the effective dose producing a 10% increased incidence of response), and determining the upper confidence limit on the model.

Chemical casualty estimating guidelines (CCEGs). Media-specific chemical concentrations expected to cause health impairments sufficient to reduce unit strength (i.e., pose a medical threat). The CCEGs are used to evaluate course-of-action options that are expected to involve chemical exposures.

Cumulative risk. Likelihood of occurrence of an adverse health effect from exposure to multiple chemicals that have common modes of toxicity from all routes and pathways.

Deployment. Unless specifically defined differently by the commander/leader responsible for the mission at hand, a deployment is defined as a troop movement resulting from a JCS/Unified Command deployment order to a land-based location outside the Continental United States that does not have a permanent U.S. Medical Treatment Facility (i.e., funded by the Defense Health Program) that lasts 30 or more consecutive days (U.S. Department of the Army, HQDA Ltr 1-0-1, 27 June 2001).

Force health protection. A unified and comprehensive strategy that aggressively promotes a health and fit force and provides full protection from all potential health hazards throughout the deployment process. Its major ingredients include healthy and fit force promotion, casualty and injury prevention, and casualty care and management (Department of the Army, HQDA Ltr 1-0-1, 27 June 2001).

Health threat. Refers to an individual soldier's health. It includes hereditary conditions that manifest themselves in adulthood, individual exposure to an industrial chemical or toxin where others are not exposed, or other injuries and traumas that affect an individual's health rather than the health of the unit (FM 4-02.17, Department of the Army, 28 August 2000).

Long-term exposure. Exposure to a toxicant or health threat with a maximum duration of one year (Department of the Army, HQDA Ltr 1-0-1, 27 June 2001).

Long-term health effect. A health effect, usually adverse, that manifests itself a significant period of time (months or years) after the causative event (i.e., exposure to a toxicant). This term is also used to describe a health effect that persists for a relatively long period of time (months or years) (Department of the Army, HQDA Ltr 1-0-1, 27 June 2001).

Medical threat. A subset of health threats that have the potential to degrade a unit's combat (or mission) effectiveness. Is defined as "a collective term used to designate all potential or continuing enemy actions and environmental situations that could adversely affect the combat effectiveness of friendly forces, to include wounds, injuries, or sickness incurred while engaged in a joint operation" (Joint Publication 4-02, *Doctrine for Health Service Support in Joint Operations*. 26 April 1995).

Military exposure guideline. An estimated chemical concentration above which certain types of adverse health effects might begin to occur in individuals within the exposed population after a continuous, single exposure of specified duration.

Occupational and environmental health. Human health issues impacted by hazardous materials, agents, organisms, or conditions found in a specific work environment or in the natural environment (Department of the Army, HQDA Ltr 1-0-1, 27 June 2001).

Occupational and environmental health threats. Threats to health of personnel and military readiness created by exposures to hazardous agents contained in or produced by weapons systems, as well as exposures to environmental contamination or toxic industrial materials (Department of the Army, HQDA Ltr 1-0-1, 27 June 2001).

Short-term exposure. Exposure to a toxicant or health threat with a maximum duration of two weeks (Department of the Army, HQDA Ltr 1-0-1, 27 June 2001).

Short-term health effect. A health effect, usually adverse, that manifests itself shortly after the causative event (i.e., an exposure to a toxicant). This term is also used to describe an adverse health effect that persists for a relatively short period of time before subsiding completely (Department of the Army, HQDA Ltr 1-0-1, 27 June 2001).